GUIDELINES FOR

PREPARING PROCESS EQUIPMENT FOR MAINTENANCE AND RETURN TO SERVICE

PUBLICATIONS AVAILABLE FROM THE CENTER FOR CHEMICAL PROCESS SAFETY OF THE AMERICAN INSTITUTE OF CHEMICAL ENGINEERS

This book is one in a series of process safety guidelines and concept books published by the Center for Chemical Process Safety (CCPS). Refer to www.wiley.com/go/ccps for full list of titles in this series.

Guidelines For
Preparing Process Equipment For Maintenance And Return To Service

CENTER FOR CHEMICAL PROCESS SAFETY

of the

AMERICAN INSTITUTE OF CHEMICAL ENGINEERS

120 Wall Street, 23rd Floor • New York, NY 10005

An AIChE Technology Alliance

CCPS Center for Chemical Process Safety

WILEY

Published by John Wiley & Sons, Inc., Hoboken, New Jersey.
Published simultaneously in Canada.

For general information on our other products and services or for technical support, please contact our Customer Care Department within the United States at (800) 762-2974, outside the United States at (317) 572-3993 or fax (317) 572-4002.

Wiley also publishes its books in a variety of electronic formats. Some content that appears in print may not be available in electronic formats. For more information about Wiley products, visit our web site at www.wiley.com.

Library of Congress Cataloging-in-Publication Data Applied for:

Hardback ISBN: 9781394237890

Cover Images: Silhouette, oil refinery © manyx31/iStock.com, Stainless steel © Creativ Studio Heinemann/Getty Images, Dow Chemical Operations, Stade, Germany/ Courtesy of The Dow Chemical Company

SKY10099113_022825

CONTENTS

Contents .... *vii*
List of Figures .... *xii*
List of Tables .... *xiii*
List of Case Studies .... *xiv*
Acronyms and Abbreviations .... *xvi*
Glossary .... *xix*
Acknowledgments .... *xxvii*
Dedication .... *xxix*
Foreword .... *xxxi*
Preface .... *xxxii*

1 INTRODUCTION 1

1.1 Purpose and Scope of the Book 1
1.2 Key Definitions Relevant to Preparing Equipment for Maintenance and Return to Service 3
1.3 Overview of the Main Focus Areas 3

2 REGULATION AND PROCESS SAFETY 5

2.1 Regulatory Aspects 5
2.2 Company Standards and Safe Systems of Work 6
2.3 Process Safety and Personal Safety 8

3 UNDERSTANDING THE HAZARDS 11

3.1 Information Requirements-Process Safety Information 11
3.2 Equipment Hazards 12
3.3 Energy Hazards 17
3.4 Simultaneous Operations 24
3.5 Human Factors 24
3.6 Frequently Encountered Issues when Preparing Equipment for Isolation 25
3.7 Summary 26

4 PLANNING FOR ISOLATION AND DECONTAMINATION 27

4.1 Plan Considerations 27
4.2 Risk Assessment 29
4.3 Isolation/Decontamination – Which Comes First? 35
4.4 Planning The Energy Isolation 36
4.5 Planning for Decontamination and Decommissioning 39
4.6 Disposal of Chemicals from Decontaminated Processes 43
4.7 Planning for Return to Service 43
4.8 Changing Plans 46

5 THE DECONTAMINATION PLAN 47

5.1 Documented Plan 47
5.2 Phase of Decontamination 49
5.3 Managing the Decontamination Hazards and Risks 50
5.4 Training for Decontamination 53
5.5 Difficult Decontaminations 55
5.6 Moving Decontaminated Equipment 57
5.7 Checklists and Equipment Identification 57

6 EXECUTING THE DECONTAMINATION PLAN 59

6.1 Sharing the Plan 59
6.2 Tracking the Plan 60
6.3 Completing and Documenting Execution of the Plan 63
6.4 Complex Decontamination Plans 64
6.5 Equipment Readiness for Energy Isolation 66
6.6 Summary 66

7 ISOLATION AND DECONTAMINATION PLAN EXAMPLES 69

7.1 Isolating/Decontaminating Lines with Toxic Materials using "Double Block and Bleed" 69
7.2 Isolation/Decontamination for Vessel Entry 73
7.3 Isolation for Longer Term/Mothballing 77
7.4 Isolating/Decontaminating Butyl Lithium Storage Systems 78
7.5 Example – Isolating/Decontaminating a Toluene Diisocyanate Reactor Train 85
7.6 Summary 92

8 OVERVIEW OF ENERGY ISOLATION 93

8.1 Degrees of Isolation 93
8.2 Human Factors in Energy Isolation 95
8.3 Designing for Energy Isolation 96
8.4 Locking and Labelling 98
8.5 Documenting Energy Isolation 99
8.6 Chemical Process Isolation 101
8.7 Electrical Isolation 101
8.8 Mechanical, Potential and Natural Energy Sources 103
8.9 Nuclear Radiation 103
8.10 Summary 104

9 WRITTEN ENERGY ISOLATION PLANS (EIP) 105

9.1 De-isolation, Testing, and Restart Considerations 106
9.2 Risk Level 106
9.3 Documentation - Safe Work Permit and EIP 109
9.4 Piping and Instrument Diagram (P&ID) Symbols for Process Isolation 110
9.5 Forms and Types of Energy Isolation 112
9.6 Lock – Tag - Try 114
9.7 Final Verification Before Release to Work 114
9.8 First Line Break 116
9.9 Monitoring Isolations 117
9.10 Summary 118

10 SPECIFIC ISOLATION REQUIREMENTS 119

10.1 Process/Instrument Isolation 119
10.2 Electrical Isolation 120
10.3 Mechanical and Potential Energy and Natural Energy Isolation 122
10.4 Nuclear Radiation 123
10.5 Boundary Isolation 124
10.6 Specialized Isolation Techniques 124
10.7 Field Verification and Changes to Isolation 125
10.8 Summary 125

11 REVERSING THE ISOLATION PLAN 127

11.1 Job Completion and Transfer of Equipment Ownership 127
11.2 De-isolation Procedure 127
11.3 Process De-isolation 130
11.4 Vents and Drains 132
11.5 De-isolation of Other Sources of Energy 132
11.6 Cleanliness 132
11.7 Bolts, Flanges, and Gaskets 133
11.8 Changes to Isolation Boundary 133
11.9 Summary 133

12 RECOMMISSIONING 135

12.1 Checks for Debris 137
12.2 Cleaning and Flushing 138
12.3 Purging 138
12.4 Leak/Pressure Testing 140
12.5 Functional Checks 144
12.6 Walk the Line (WTL) 146
12.7 Delayed Return to Service 147

13 PRE-STARTUP SAFETY REVIEW 149

13.1 PSSR and Returning Equipment to Service 149
13.2 PSSR Overview 152
13.3 PSSR Example: Short Form 152
13.4 Summary 156

14 EQUIPMENT STARTUP 157

14.1 Participation and Communication 157
14.2 Specific Startup Issues 158
14.3 Startup of Mothballed Equipment 161

15 CONCLUSION 165

APPENDIX A: WORK PERMIT 167

APPENDIX B: WALK THE LINE 171

APPENDIX C: DECONTAMINATION CERTIFICATE 177

APPENDIX D: LONG FORM FOR HIGHER RISK/ COMPLEX PSSR 179

REFERENCES 189

INDEX 197

LIST OF FIGURES

Figure 1-1 Flowchart for Equipment Maintenance and Return to Service 2
Figure 4-1 First Line Break Against Single Valve 35
Figure 4-2 Purge Valve Fitted to Blind Flange 45
Figure 6-1 Fire at Husky Refinery 61
Figure 7-1 Isolation Design for a Piping System Containing a Toxic Liquid Chemical 70
Figure 7-2 Vessel Isolation Schematic for Confined Space Entry 74
Figure 7-3 Diagram of TDI CSTR System 86
Figure 8-1 Spectacle Blind Shown in Open Position 96
Figure 8-2 Fast Acting Blind Valves 97
Figure 8-3 Electrical Isolator Lock 97
Figure 8-4 Multi-hasp and Lock Box 99
Figure 9-1 Key to Typical P&ID Symbols for Energy Isolation 111
Figure 9-2 Process Isolation with Increasing Integrity of Isolation 113
Figure 10-1 Electrical Switchbox Used for Training 121
Figure 11-1 Remains of Piper Alpha on the Morning Following the Explosions and Fires 128
Figure 11-2 De-isolation with Unproven Absence of Energy Source 130
Figure 11-3 De-isolation with Ability to Prove Absence of Energy Source 131
Figure 11-4 De-isolation with Bleed Valve Allowing to Check for Absence of Energy Source 131
Figure 13-1 High Level PSSR Workflow Chart 150
Figure A-1 Safe Work Permit 168
Figure B-1 Conduct of Operations Management System Model 172

LIST OF TABLES

Table 3-1 Equipment Hazard Examples 14
Table 3-2 Energy Hazards 17
Table 3-3 Chemical Hazard Examples 18
Table 3-4 Frequently Encountered Issues to be Considered 25
Table 4-1 Example Job Safety Analysis Worksheet 32
Table 4-2 Planning for Return to Service 44
Table 5-1 Key Components of Decontamination Plan 48
Table 6-1 Decontamination Checklist 66
Table 7-1 Isolation and Decontamination Steps for Hazardous Material 71
Table 7-2 Decontamination for Vessel Entry 74
Table 7-3 Isolation and Decontamination Checklist 82
Table 7-4 Return to Service Checklist 84
Table 7-5 Outline Standard Operating Procedure for Isolation and Decontamination 89
Table 8-1 Relative Security Level of Isolation Types for Different Energy Sources 95
Table 8-2 Typical Energy Isolation Documentation and Respective Usage 100
Table 13-1 Global Examples of Pre-Startup Safety Review Related Documents 151
Table 13-2 Example Short Form for Lower Risk/Simple PSSR 153
Table C-1 Decontamination Certificate 178
Table D-1 Example Long Form for Higher Risk/Complex PSSR 181

LIST OF CASE STUDIES

Case Study 2-1 — Replacing a Sensor: New Design Leads to a Fire! 9
Case Study 2-2 — No Isolation: Possible Radiation Danger! 10
Case Study 4-1 — Breaking into the Wrong Line 29
Case Study 4-2 — Chemical Exposure 34
Case Study 4-3 — Isolation on an Oil Refinery 38
Case Study 4-4 — Isolated but Not Fully Decommissioned - 1 40
Case Study 4-5 — Left a Valve Dead-leg – Not Fully Decommissioned - 2 41
Case Study 4-6 — Isolated Longer than Intended 42
Case Study 5-1 — Toxic Break-in 49
Case Study 5-2 — Refinery Storage Tank Explosion 52
Case Study 5-3 — Solids Plugged in a Thermosyphon Reboiler 53
Case Study 5-4 — The Tank That Was Not Isolated 54
Case Study 5-5 — Pyrophoric Ignition of Structured Packing in a Crude Vacuum Column 56
Case Study 6-1 — Explosion/Fires at Refinery During Maintenance Turnaround 61
Case Study 6-2 — Decontamination of a Toxic Pipeline 65
Case Study 8-1 — Electrical Low Voltage Isolation 102
Case Study 9-1 — Lack of Energy Isolation 107
Case Study 9-2 — Inadequate Isolation 109
Case Study 9-3 — Filter Cartridges Change Out 110
Case Study 9-4 — Flash Fire Caused by Inadequate Isolation 114
Case Study 9-5 — Safety Relief Valve LOTO Compromised 118
Case Study 10-1 — Vapor Cloud Explosion at Phillips 66 Facility 120
Case Study 11-1 — Piper Alpha 128
Case Study 12-1 — Vent Line Blockage 138
Case Study 12-2 — Vessel Failure During Pneumatic Pressure Test 141
Case Study 12-3 — Thermal Oil System Boil Over and Fire 143
Case Study 12-4 — Process Safety Scrubber: Testing is Key! 145
Case Study 12-5 — Spare Reboiler Explosion 148

Case Study 13-1 — Oil Pipe Rupture after Commissioning 154
Case Study 13-2 — Fire Heater Explosion after Restart 155
Case Study 14-1 — Kleen Energy Natural Gas Explosion 160
Case Study 14-2 — ARCO Channelview Compressor Explosion 162
Case Study 15-1 — Solvent Release from Bleed on a 'Double Block and Bleed' Valve 166

ACRONYMS AND ABBREVIATIONS

ACOP	Approved Codes of Practice
ACC	American Chemistry Council
AIChE	American Institute of Chemical Engineers
BLEVE	Boiling Liquid Expanding Vapor Explosion
CCPS	Center for Chemical Process Safety
CFR	Code of Federal Regulations
CISCC	Chloride Induced Stress Corrosion Cracking
COD	Consequence of Deviation
COMAH	Control of Major Accident Hazards
CSB	Chemical Safety Board (US)
CSTR	Continuously Stirred Tank Reactor
DB&B	Double Block and Bleed
DCS	Distributed Control System
DSEAR	Dangerous Substances and Explosive Atmospheres Regulations
EA	Environment Agency
EIP	Energy Isolation Plan
EPA	Environmental Protection Agency
FCC	Fluid Catalytic Cracking
FMEA	Failure Mode and Effects Analysis
FMECA	Failure Modes, Effect and Criticality Analysis
GMP	Good Management Practice
HASAWA	Health and Safety at Work Act
HAZOP	Hazard and Operability Study
HDPE	High Density Polyethylene
HIRA	Hazard Identification and Risk Analysis
HRSG	Heat Recovery Steam Generator
HSE	Health and Safety Executive
HSG	Health and Safety Guidance
ICI	Imperial Chemical Industries
ILO	International Labor Office
IOGP	International Association of Oil and Gas Producers
ISO	International Organization for Standardization
ITPM	Inspection, Testing, and Preventive Maintenance
JSA	Job Safety Analysis
KO	Knock Out [drum]
LEL	Lower Explosive Limit

LNG	Liquefied Natural Gas
LOPA	Layer of Protection Analysis
LOPC	Loss of Primary Containment
LOTO	Lock-Out Tag-Out
LPG	Liquefied Petroleum Gas
MAWP	Maximum Allowable Working Pressure
MIC	Microbiological Introduced Corrosion
MOC	Management of Change
NDE	Non-Destructive Examination
NFPA	National Fire Protection Association
NORM	Naturally Occurring Radioactive Material
NRC	Nuclear Regulatory Commission
OECD	Organization for Economic Cooperation and Development
OEM	Original Equipment Manufacturer
ONR	Office for Nuclear Radiation
OSHA	Occupational Safety and Health Administration
P&IDs	Piping & Instrumentation Diagrams
PEI	Positive Equipment Identification
PFD	Process Flow Diagram
PHA	Process Hazard Analysis
PIS	Pyrophoric Iron Sulfide
POSM	Propylene Oxide/Styrene Monomer
PPE	Personal Protective Equipment
PRD	Pressure Relief Devices
PSE	Process Safety Event
PSI	Process Safety Information
PSM	Process Safety Management
PSSR	Pre-Startup Safety Review
PSV	Pressure Safety Valve
PTW	Permit to Work
RBPS	Risk Based Process Safety
RMP	Risk Management Program
SCBA	Self-Contained Breathing Apparatus
SCTA	Safety Critical Task Analysis
SDS	Safety Data Sheets
SEPA	Scottish Environment Protection Agency
SIMOPS	Simultaneous Operations
SME	Subject Matter Expert
SOL	Safe Operating Limit
SOP	Standard Operation Procedure
SSOW	Safe Systems of Work

SWA	Stop Work Authority
TIH	Toxic Inhalation Hazard
TLV	Threshold Limit Values
UNEP	United Nations Environment Programme
UV	Ultraviolet
VRU	Vapor Recovery Unit
WTL	Walk the Line

GLOSSARY

Term	Definition
Air-gapping	Removal of pipe or spool piece between process and work area (Physical disconnection)
Barrier	A control measure or grouping of control elements that on its own can prevent a threat developing into a top event (prevention barrier) or can mitigate the consequences of a top event once it has occurred (mitigation barrier). A barrier must be effective, independent, and auditable (CCPS 2019). For the subject matter of this book, this may include a closed valve, a blinded or blanked line, electrical lock-out or a safety procedure relating to a check before a line is opened.
Blank Flange or Blanking Plate	See Blind Flange.
Blind/Line Blind	A blind is a solid plate, usually fitted with a tab to aid identification, handling, and removal, inserted between two flanges to achieve a form of positive isolation. Also referred to as a Spade. See Figure 4-1. Note: Not considered as secure as physical disconnection due to potential hidden corrosion of blind.
Blind Flange	Circular plate bolted to flange at end of line and fitted with gasket to achieve positive isolation (physical disconnection). Also referred to as a Blank Flange.
"Bump Start"	See Try-Out.
Cap	A female screwed fitting (typically small bore) that achieves positive isolation (physical disconnection) by screwing onto an externally threaded section at the end of a pipe, on a drain/bleed valve, or on a piece of equipment. See Plug.

Term	Definition
Car Seal	A metal or plastic cable attached to a piece of equipment (typically a valve) that shows it should not be moved without proper authorization. While it is intended to be strong enough to prevent operation of the valve, it is not as robust as, for example, a padlock. See LOTO.
Checklist	A list of items or steps requiring verification of completion; typically, a procedure format in which each critical step is marked off (or otherwise acknowledged/verified) as it is performed (CCPS 2019).
Decontamination	Process of removal of substance(s) from equipment so that it can be worked on/disassembled without presenting a safety risk to personnel or the environment.
Decommissioning	Completely de-inventorying all materials from a process unit and permanently removing the unit from service (CCPS 2019).
De-isolation	Removing the items/devices that are excluding the sources of energy at defined points.
Design Pressure	The maximum pressure that the equipment is designed to contain, used by the mechanical designer to size the pressure-containing parts. This will typically be at or above the maximum operating pressure of the equipment.
Double Block and Bleed (DB&B)	Closure of two valves with a valved vent or drain between them. This configuration allows for the release of trapped pressure, fluid, or gas between the two blocked valves, providing an added assurance of isolation.
Energy Source	In the context of isolation from an energy source for maintenance, this could include electricity, hazardous chemicals, water, steam, pneumatic pressure, hydraulic pressure, temperature (high and low), mechanical or potential energy, radioactive materials, among others.

Term	Definition
Energy Isolation Plan (EIP)	A document that shows steps to be taken and verifications required to prevent the accidental release of energy when work is to be conducted on a process, plant, or equipment.
First Line Break	Initial, controlled opening of a line or vessel that may contain a source of energy.
Flammable	Subject to combustion if exposed to an ignition source. Per NFPA: Capable of burning with a flame. Dusts, gases, and vapors if in a confined area, can potentially explode.
Hazard Identification and Risk Analysis (HIRA)	A collective term that encompasses all activities involved in identifying hazards and evaluating risk at facilities, throughout their life cycle, to make certain that risks to employees, the public, or the environment are consistently controlled within the organization's risk tolerance (CCPS 2019).
Inspection, Testing and Preventive Maintenance (ITPM)	Scheduled proactive maintenance activities intended to (1) assess the current condition and/or rate of degradation of equipment, (2) test the operation/functionality of equipment, and/or (3) prevent equipment failure by restoring equipment condition.
Isolation	Means of preventing a release of energy beyond a defined point.
Job Safety Analysis (JSA)	A procedure that systematically identifies: 1) job steps, 2) specific hazards associated with each job step, and 3) safe job procedures associated with each step to minimize accident potential. Also called job hazard analysis (CCPS 2019).
Lock-Out-Tag-Out (LOTO)	A safe work practice in which energy sources are positively blocked away from a segment of a process with a locking mechanism and visibly tagged as such to help ensure worker safety during maintenance and some operations tasks (CCPS 2019).
Leaks by (of valve)	Fluid flowing from the high pressure side through the internals of a valve to the lower pressure side, also referred to as "passing".

Term	Definition
Leak Test	Pressurizing equipment (typically up to maximum operating pressure) using a liquid or gas (e.g., nitrogen, helium), while observing for leaks; this may include a soapy water test, "sniffer", or monitoring a drop in pressure.
Life Saving Rules	A series of key instructions that cannot be broken, which target some of the key causes of injuries and fatalities in the process industries.
Long-term Isolation	Prolonged isolation of equipment during an extended timeframe (e.g., over 90-days). This often involves comprehensive measures and controls to maintain the isolation status over a longer timeframe compared to standard LOTO procedures, This may include additional measures (preservation plans), operator checks, and MOC.
Loss of Primary Containment (LOPC)	An unplanned or uncontrolled release of material from primary containment, including non-toxic and non-flammable materials (e.g., steam, hot condensate, nitrogen, compressed CO_2 or compressed air) (CCPS 2019).
Management of Change (MOC)	A management system to identify, review, and approve all modifications to equipment, procedures, raw materials, and processing conditions, other than replacement in kind, prior to implementation to help ensure that changes to processes are properly analyzed (for example, for potential adverse impacts), documented, and communicated to employees affected (CCPS 2019).
Maximum Allowable Working Pressure (MAWP)	Maximum pressure at specific temperature at which the vessel can safely function.
Mothballed	Taken out of service, usually disconnected (see Long-term Isolation) and preserved, with the intention of re-using/reinstating the equipment/process at a later date.

Term	Definition
Out of service	Taken offline, but not necessarily to be prepared for maintenance.
Operational Discipline (OD)	The performance of all tasks correctly every time. Individuals demonstrate their commitment to process safety through OD, executing the organization's Conduct of Operation (COO) RBPS Element each and every day (CCPS 2019).
Piping and Instrumentation Diagrams (P&IDs)	The key source of data used in preparation for process equipment isolations, providing a schematic drawing of all pipes and instruments and vessels/equipment. It must be up to date.
Plug	A male screwed fitting (typically small bore) that achieves positive isolation (physical disconnection) by screwing inside a threaded section at the end of a pipe, on a drain/bleed valve, or on a piece of equipment. See Cap.
Positive Isolation	The most secure isolation. Includes inserting blind or air-gapping the pipe from the isolated equipment.
Pressure Safety Valve (PSV)	A pressure relief device which is designed to reclose and prevent the further flow of fluid after normal conditions have been restored.
Pressure Test	Pressurizing sealed equipment with a fluid (vapor = pneumatic or liquid = hydrostatic) to a designated test pressure (from design pressure up to test pressure) to check for mechanical integrity of equipment. May include monitoring drop in pressure.
Procedures	Written step-by-step instructions and associated information (cautions, notes, warnings) that describe how to safely perform a task (CCPS 2019).
Pre-Startup Safety Review (PSSR)	A systematic and thorough check of a process prior to the introduction of a highly hazardous chemical to a process. The PSSR must confirm the following: Construction and equipment are in accordance with design specifications; Safety, operating, maintenance, and emergency procedures are in place and are adequate; A process hazard analysis has been performed for new facilities and recommendations

Term	Definition
	have been resolved or implemented before startup, and modified facilities meet the management of change requirements; and training of each employee involved in operating a process has been completed (CCPS 2019).
Process Safety	A disciplined framework for managing the integrity of operating systems and processes handling hazardous substances by applying good design principles, engineering, and operating practices. Note: Process Safety focuses on efforts to reduce process safety risks associated with processes handling hazardous materials and energies. Process Safety efforts help reduce the frequency and consequences of potential incidents. These incidents include toxic or flammable material releases (loss events), resulting in toxic effects, fires, or explosions. The incident impact includes harm to people (injuries, fatalities), harm to the environment, property damage, production losses, and adverse business publicity (CCPS 2019).
Process Safety Management (PSM)	A management system that is focused on prevention of, preparedness for, mitigation of, response to, and restoration from catastrophic releases of chemicals or energy from a process associated with a facility(CCPS 2019). In the USA, OSHA (Occupational Safety and Health Administration) has issued regulatory standard 29 CFR 1910.119; Process Safety Management of Highly Hazardous Chemicals (OSHA 2023).
Recommissioning	Activities involved from the point at which maintenance, repairs, or modifications are complete, including testing, up to the re-introduction of chemicals to the process.
Purging	Process of removing one fluid (gas or liquid) and replacing it with another, usually less hazardous gas.
Pyrophoric	A chemical with an autoignition temperature in air at or below 130 °F (54.4 °C) (NFPA 2022).
Readiness Review	A work activity that occurs prior to initial startup or restarting a process unit to verify that the condition of process equipment and safety systems, the status of

Term	Definition
	limiting conditions for operations, and in some cases, the training and qualification status of personnel conform to predefined conditions (CCPS 2019).
Return to Service	The final act of bringing a facility or equipment back online after recommissioning, by reintroducing chemicals and other sources of energy, and re-establishing normal operations.
Safe Operating Limits	Limits established for critical process parameters, such as temperature, pressure, level, flow, or concentration, based on a combination of equipment design limits and the dynamics of the process.
Safe Work Practices	An integrated set of policies, procedures, permits, and other systems that are designed to manage risks associated with non-routine activities such as performing hot work, opening process vessels or lines, or entering a confined space (CCPS 2019). Also referred to as Safe System of Work (SSOW).
Safety Critical Task Analysis	An analysis to identify activities where there is the potential for human error that could lead to an incident, allowing appropriate mitigations to be put in place.
Spade	See Blind.
Spectacle Blind/Spectacle Blank	A single plate comprising both a blind and an open flange ring, in a "figure 8" pattern, which is installed between flanges and can be rotated between a line-open and line-closed position. See Figure 8-1.
Simultaneous Operations (SIMOPS)	Situations where two or more operations or activities occur at the same time and in the same place or system in a facility.
Subject Matter Expert (SME)	An individual with detailed and specialized knowledge in a specific area that could include equipment, processes, techniques, and/or procedures.
Stop Work Authority (SWA)	Authority granted to personnel, to stop work when an unsafe condition or act is observed.
Test Pressure	Pressure at which equipment is tested, that depends on situation and standards that apply (e.g., for ASME

Term	Definition
	B31.3—test pressure for piping is usually 1.5 × design pressure for a hydrotest and 1.1 × design pressure for a pneumatic test.)
Toxic	Harmful to people, animals, or the environment.
Transient Operations	A period when the process is in transition between one operating state and another.
Try-Out	Physical attempt to operate the isolated equipment that has been subjected to LOTO, in order to assess its effectiveness. Sometimes referred to as "Bump Start".
Valve Isolation	Single Block: Using a single valve to block flow in a system. This is the most basic type of isolation and may not be deemed sufficient for certain high-risk operations.
Walk the Line (WTL)	The practice of physically checking in the field that all equipment, including valves, joints, instruments, isolations, de-isolations are as they should be, to avoid line-up errors and potential loss of containment, before energy is restored to the system. These physical checks form part of a conduct of operations management system model consisting of operational continuity, operational discipline, and operational readiness tools that address errors and mistakes. See 12.6 and Appendix B.
Work Permit	A formal document that includes a description of the planned activities, the hazards, risks, and mitigations in place and residual risks that may be encountered. Requires signatures from suitably qualified workers both for the permit issuer and the recipient, and on completion of the task. Also referred to as a Safe Work Permit and Permit to Work.

ACKNOWLEDGMENTS

The American Institute of Chemical Engineers (AIChE) and the Center for Chemical Process Safety (CCPS) express their appreciation and gratitude to all members of the *Guidelines for Preparing Process Equipment for Maintenance and Return to Service* Subcommittee for their generous efforts in the development and preparation of this important guideline. CCPS also wishes to thank the subcommittee members' respective companies for supporting their involvement during the different phases in this project.

Subcommittee Members:

Jerry Forest, Chair	Jerry Forest, LLC
Pete Lodal, CCPS Consultant	D&H Process Safety
Steve Courson	Wacker Chemical
Lisa Faris	Covestro
Amy Gay	OxyChem
Peter Herena	CF Industries
Andy Hiester	Eastman Chemical
Luke Kittmer	CF Industries
James Martin	Corteva
Becky McDonald	ABS
Trey Morrison	Exponent
Miguel Rizo	Contra Costa County
William Vassalo	Chevron

The CCPS book committee wishes to express their appreciation to BakerRisk, the contractor for developing the manuscript for this book, led by Roger Stokes and Larry Bowler. Phyllis Whiteaker was BakerRisk document editor and James Birch provided the section on butyl lithium. Jennifer Bitz, CCPS, provided final review and editing. Before publication, all CCPS guidelines are subjected to a peer review process. CCPS gratefully acknowledges the thoughtful comments and suggestions of the peer reviewers. Their work enhanced the accuracy and clarity of this guideline.

Although the peer reviewers provided comments and suggestions, they were not asked to endorse this guideline and did not review the final manuscript before its release.

Peer Reviewers:

Kenan Stevick	KPS, Inc.
John Herber	Process Hazard Management Services, LLC
Nicole Loontjens	Americas Styrenics LLC
Bruce Vaughen	CCPS

DEDICATION

Seeing the grass bleached near your house can be a fascinating experience for a 9-year-old. But the release from fire at the chlorinated dry bleach manufacturing facility that made it happen, left a lifelong impression on Louisa Nara. It also led her to become a Chemical Engineer and subsequently focus on Process Safety as a career.

Louisa holds a Bachelor of Science in Chemical Engineering from West Virginia University, and Master of Science in Environmental Engineering from Villanova University. During her career, she worked at various chemical and specialty chemical companies.

Imagine, if you will, a person who has been, throughout the course of a career a 1) Process Safety and Crisis Manager for a plant, 2) successful author, 3) legal department consultant for a major international company, 4) training developer, 5) instructor, and 6) unfailingly energetic advocate for process safety. I think you would call this individual an extraordinary person. We call her – Louisa Nara.

Louisa has accomplished all of this in her remarkable career. She is indeed an extraordinary individual. More importantly, she is a great friend, and role model for the process safety community worldwide.

Louisa Nara was recognized as one of the Legends of Process Safety in 2023. She is a CCPS certified Process Safety professional (CCPSC), a recognition as one of the world leading subject matter experts. She is also an AIChE and CCPS Fellow.

Benefiting from great mentors early in her career, Louisa continued this tradition inspiring countless students and professionals with her optimistic demeanor and unwavering support. Even after retiring as CCPS Global Technical Director in 2023, she continues to teach and volunteer for various CCPS initiatives.

This book is dedicated to Louisa Nara, who embodies the CCPS vision of a World without Process Safety Incidents. Her passion, achievements, and tireless dedication have not only advanced the field but have also touched the lives of many.

Dr. Anil Gokhale, PE
Chief Operating Officer, CCPS

FOREWORD

CCPS, through the efforts of its Technical Steering Committee (TSC) and Planning Subcommittee are constantly seeking out topics in process safety that, for one reason or another, have not been fully recognized and documented in the literature. Preparing Process Equipment for Maintenance and Return to Service, the topic of this Guideline Book, was one such area in need of improvement. Indeed, the fact that more than 40 case studies have been included in the text demonstrates the need for the information contained in this Guideline.

Many of the safe work practices required for successful preparation for maintenance and return to service are already in existence with a heavy emphasis on energy isolation and operational readiness. This text combines each of these elements with an overall hazard identification and risk analysis emphasis providing a complete view of this phase of operation. This guideline brings together these elements with hazard evaluation steps commensurate with process hazards interspersed with development of detailed, step-by-step decontamination, de-energization, equipment removal and reinstallation, re-introduction of process materials, and returning equipment to service viewed as a complete process to prevent process safety incidents during these activities.

In process safety, there is no more emotionally difficult task than a fatality investigation. Both of us have had the unfortunate opportunity to conduct such an investigation as a direct result of failure to perform the return to service tasks effectively. Hence, our motivation to drive this project to a successful conclusion. We have assembled a team with several hundred years of manufacturing experience to identify and share industry practices in addition to our own experiences. It is our sincere hope that, using the tools, techniques, and procedures in this Guideline, none of you will have to go through what we did.

Jerry Forest
Jerry Forest, LLC
Lake Charles, LA

Peter Lodal
D&H Process Safety
Kingsport, TN

PREFACE

This book is intended to help prevent process safety, personal safety, and environmental events that can occur when Preparing Process Equipment for Maintenance and Return to Service.

The key focus of the book is understanding hazards and managing risk associated with the transient operating phases of preparing equipment for maintenance and other downed states, in particular the decontamination phase, energy isolation, and equipment readiness for return to service. Multiple significant process safety events have occurred due to a failure to understand and manage these risks, and various case studies are placed throughout the book in order to illustrate the concepts as they are introduced.

The target audience is global, primarily operations and maintenance management/supervisors, maintenance/turnaround planners, overhaul teams, recommissioning teams, fitters, control room operators, laboratory personnel (testing), and process safety engineers who are involved with and responsible for process operations.

The American Institute of Chemical Engineers (AIChE) has been closely involved with process safety and loss control issues in the chemical, petrochemical, and allied industries for more than four decades. AIChE publications and symposia have become information resources for those devoted to process safety and environmental protection.

AIChE created the Center for Chemical Process Safety (CCPS) in 1985 after significant chemical disasters in Mexico City, Mexico, and Bhopal, India. The CCPS is chartered to develop and disseminate technical information for use in the prevention of major chemical accidents. The center is supported by more than 200 chemical process industry sponsors that provide the necessary funding and professional guidance to its technical committees.

The major product of CCPS activities has been a series of guidelines to assist those implementing various elements of a process safety and risk management system. This book is part of that series.

CCPS strongly encourages companies around the globe to adopt and implement the recommendations contained within this book.

1 INTRODUCTION

1.1 PURPOSE AND SCOPE OF THE BOOK

The purpose of the book is to help prevent process safety, personal safety, and environmental events that are caused, either directly or indirectly, by activities being conducted on equipment or processes when preparing process equipment for maintenance and return to service. The concepts and practices that are discussed within this book could also be applied for temporary or permanent decommissioning of process equipment.

Numerous steps and multiple groups of personnel from a variety of functions and backgrounds are involved in these activities. The overall responsibility for setting policy and standards to provide safe systems of isolation for maintenance and return to service clearly lies with senior leadership. Although the target audience for this book includes senior management, it is primarily directed at the operations and maintenance management/supervisors, maintenance/turnaround planners, equipment overhaul teams, recommissioning teams, technicians, control room operators, laboratory personnel (sampling and testing), and process safety engineers who are responsible for carrying out such activities.

Because there are multiple examples of significant process safety events that have occurred where weaknesses in the planning or execution of shutdown/restoring service activities were causal factors, actual case studies are placed throughout the book in order to illustrate the concepts as they are introduced. This book is not meant to replace a company's existing process safety programs, but to add understanding and demonstrate the value of having a complementary Preparing Process Equipment for Maintenance and Return to Service management practice in place.

> *The use of the words "must" and "should" throughout this guideline book is intended to convey the relative importance of the advice presented and does not imply a legal requirement, but rather provides a guide on what is considered best practice.*

A simple flow chart is provided in Figure 1-1 showing recommended steps in successfully shutting down and returning equipment to service and the applicable chapters in the book that discuss each activity.

Identifying and Understanding the Hazards – Chapter 3
Fundamental requirement to establish the potential risks associated with the activities associated with equipment shutdown and return to service, including process safety information, equipment hazards and energy hazards. Identification/verification of skill sets required.

↓

Risk Assessment – Chapter 4
Assessment of risk, based on hazards identified and risk mitigations that should be put in place, commensurate with magnitude of risk.

↓ ↑

Planning of Isolation/Decontamination Scheme – Chapter 4
Selection of the most appropriate method and degree of energy isolation and decontamination (or decontamination and then isolation), also considering the later requirements for the safety and practicality of de-isolation and return to service.

↓

Carrying out Isolation and Decontamination – Chapters 5–10)
Sharing and tracking the documented plan in a structured way, incorporating checklists and other verifications to ensure that the energy isolation and decontamination is safe and effective, with a final signoff by competent personnel. During this step, the plan must be reviewed with all affected personnel.

↓

Conducting Work Activity
Maintenance, inspection, or modification of work plan as needed.

↓

De-isolation – Chapter 11
De-isolation is not necessarily a reversal of the isolation plan. Requires checklists/signoff by competent personnel that work is complete, equipment ownership is transferred back to operations.

↓

Recommissioning and Return to Service – Chapters 12–14
Structured process of final checks before introducing energy back to the system, including leak testing and operation of equipment including safety systems. Walk the Line. For major modifications or large turnarounds, may form part of a PSSR.

Figure 1-1 Flowchart for Equipment Maintenance and Return to Service

1.2 KEY DEFINITIONS RELEVANT TO PREPARING EQUIPMENT FOR MAINTENANCE AND RETURN TO SERVICE

An overall Glossary on Page xix provides detailed explanations of the terms that relate specifically to Isolation of Equipment for Maintenance and Return to Service. A table of Acronyms and Abbreviations is provided on Page xvi.

1.3 OVERVIEW OF THE MAIN FOCUS AREAS

This book focuses on the key hazards and risks that are associated with preparing process equipment for maintenance and return to service, the involvement and interactions among multiple teams, and recommended practices for conducting the various activities. The activities involve transient operations and do not typically form part of the standard operating procedures. Thus, they are subject to change, which increases the potential for errors. It is imperative that the teams and individuals understand the risks and complexities of decontamination of the plant, testing, and return to service, so that they can plan, assess, and implement protective measures that are commensurate with the risk. The key focus areas include:

- Legislation and Process Safety (Chapter 2)
- Understanding the Hazards and Risks (Chapter 3)
- Planning and Risk Assessment (Chapter 4)
- Decontamination (Chapters 5, 6, and 7)
- Energy Isolation (Chapters 8, 9, and 10)
- Reversing the Isolation (Chapter 11)
- Testing and Recommissioning (Chapter 12 and 13)
- Equipment Startup (Chapter 14)

Mothballing and subsequent return to service or decommissioning and demolition are discussed in Sections 4.5 and 13.3.3.

Other factors considered that are relevant to safe isolation include:

- Management responsibility
- Design for safe isolation
- Equipment identification
- Installed spare equipment
- Changes to isolation procedures
- Training and competency
- Human Factors

Substantial guidance already exists concerning certain stages of these activities, and guidance on Energy Isolation, including CCPS (CCPS 2018), UK HSE Guide 253 (HSE 2006), and the Energy Institute (Energy Institute 2023).

These guidelines are intended to provide a common source for all the activities associated with the preparation of process equipment for maintenance and return to service.

2 REGULATION AND PROCESS SAFETY

2.1 REGULATORY ASPECTS

Regulations in most countries require a safe system of work practices to be specified and conducted before the work commences on any system that can potentially cause harm or injury to people or the environment.

In the USA, the Code of Federal Regulations (CFR) includes CFR 1910.119 "Process Safety Management of Highly Hazardous Chemicals" - OSHA PSM (OSHA 2023) which applies to facilities that use specified hazardous chemicals above established threshold quantities. Section (f) *Operating Procedures* refers to a requirement for written procedures and instructions for various operating phases including: (f)(1)(F) *Normal shutdown* and (G) *Startup following a turnaround or after an emergency shutdown*. Section (f)(4) states:

> *The employer shall develop and implement safe work practices to provide for the control of hazards during operations such as lockout/tagout; confined space entry; opening process equipment or piping; and control over entrance into a facility by maintenance, contractor, laboratory, or other support personnel. These safe work practices shall apply to employees and contractor employees.*

Other relevant parts of the standard include requirements for *Employee Participation; Process Safety Information; Training; Process Hazard Analysis; Pre-startup Safety Review; Mechanical Integrity; Hot Work Permit;* etc. While compliance with the OSHA PSM is mandatory for certain processes in the USA, following such practices for all facilities that involve hazardous materials and processes is highly recommended.

The US Environmental Protection Agency (EPA) regulations require that a facility must develop a Risk Management Program (RMP) if it uses certain hazardous substances above specified threshold quantities. The Risk Management Plan must identify the potential effects of a chemical accident and the steps the facility will take to prevent an accident and must also spell out emergency response procedures that will be enabled should an accident occur. The *Chemical Accident Prevention Provisions* fall under CFR Title 40 *Protection of Environment*, Chapter I(C) Part 68, and Subpart C *Prevention Program (*CFR online), which includes most of the requirements specified under OSHA PSM. Again, it is good practice to establish similar procedures as part of a normal routine for all facilities that involve hazardous materials and processes. Within US OSHA's standards is a General Duty Clause that is applicable to other materials and chemicals that can pose a hazard to people and the environment.

In the UK, the legislative requirements are less prescriptive than in the US and there is a duty of care under the Health and Safety at Work etc. Act 1974 (HASAWA) (Legislation.gov.uk 1974) to ensure, so far as is reasonably practicable, the health, safety, and welfare at work of all employees. In addition, the Health and Safety Executive (HSE) issues a number of safety regulations and approved codes of practice (ACOP) and Health and Safety Guidance (HSG), that employers, staff, and other stakeholders are expected to follow, in order to comply with their duties under HASAWA. Relevant Legislation includes: *The Control of Major Accident Hazards, 2015* (HSE 2015), *Confined Spaces Regulations,* (Legislation.gov online 1997); and *Dangerous substances and explosive atmospheres regulations (DSEAR)* (HSE 2002). Relevant HSGs include: *The safe isolation of plant and equipment,* (HSE 2006); *Electricity at work: Safe working practices,* (HSE 2013); and *Guidance on permit-to-work systems: A guide for the petroleum, chemical and allied industries,* (HSE 2005).

In the European Union, the Seveso directive on the control of major accident hazards involving dangerous substances is implemented in Great Britain through the Control of Major Accident Hazards (COMAH) Regulations.

The regulatory environment in other nations is sometimes similar to the UK or US model and their specific requirements must be considered in the context of preparing equipment for maintenance and return to service.

2.2 COMPANY STANDARDS AND SAFE SYSTEMS OF WORK

Many organizations, especially those handling highly hazardous materials, have internal written practices and procedures that usually go beyond the legislative requirements, are the responsibility of senior leadership, and are reviewed and revised on a regular basis. Companies and trade organizations have adopted the principles and practices of "Life-Saving Rules" many of which apply to key stages of isolating a plant for maintenance and returning it to service afterward. For example, the International Association of Oil and Gas Producers (IOGP 2018) produced a set of rules, based on an analysis of incidents leading to fatalities. Those rules include the following, which are relevant to preparing process equipment for maintenance and return to service:

- Work Authorization – Work with a valid permit.
- Energy Isolation – Verify Isolation and zero energy before work begins.
- Hot Work – Control flammables and ignition sources.

- Line of Fire – Keep yourself and others out of the line of fire.
- Working at height – Protect yourself from a fall when working at height.
- Confined Space – Obtain authorization before entering a confined space.
- Bypassing Safety Controls – Obtain authorization before overriding or disabling safety controls.
- Safe Mechanical Lifting – Plan lifting operations and control the area.

Further examples of company "Life-Saving Rules" include:

- First line breaks – Use at least 2 barriers.
- PPE – Wear the correct PPE for the task and the substances.
- Management of Change – Follow MOC procedure.
- Operating Limits – Keep within safe operating limits.
- Plugged equipment – Beware of lines, vessels, and dead-legs that may be plugged.

Company standards that apply to isolation for maintenance and return to service will typically drive the generation of documentation and training that is not only specific to the process but may extend to the equipment level. However, it is impractical for documentation to be pre-prepared for each and every piece of equipment or process that may be introduced throughout the lifecycle of a facility; thus, understanding and managing the risks associated with specific tasks involving isolation for maintenance and return to service are required.

2.2.1 Permit to Work

Safe systems of work (SSOW) and work permits are required when conducting activities at work such as hot work, line and equipment opening, and confined space entry. The SSOW is a set of procedures and policies designed to eliminate or minimize risks to workers or the environment. A permit to work is an example of an SSOW and is a formal document verifying that a review of the safe work practices to control potentially hazardous tasks was completed. The permit to work typically incorporates a basic risk assessment and a checklist for permit issuers and receivers to identify the hazards and understand the key precautions that are required for the work to be conducted safely. The permit is issued by a competent person and signed by the issuer and the receiver.

However, for higher risk activities such as opening up process lines (break-ins), hot work, confined space entry, testing and commissioning, the permit should refer to the additional SSOWs or risk assessments and associated documents required, such as a Job Safety Analysis (JSA), as detailed under Section 4.2.1. A typical permit to work is provided in Appendix A (Occidental 2023).

2.3 PROCESS SAFETY AND PERSONAL SAFETY

This book provides guidelines to help prevent incidents that could result in a Process Safety Event (PSE) due to hazards arising from the process; typically involving loss of containment that may lead to injuries. However, personal injuries may also arise from events that are not necessarily directly due to the process. Both personal safety and process safety are considered in this book.

2.3.1 Process Safety

Process Safety is defined in the CCPS Glossary (CCPS 2019) as

> *A disciplined framework for managing the integrity of operating systems and processes handling hazardous substances by applying good design principles, engineering, and operating practices. Note: Process Safety focuses on efforts to reduce process safety risks associated with processes handling hazardous materials and energies. Process Safety efforts help reduce the frequency and consequences of potential incidents. These incidents include toxic or flammable material releases (loss events), resulting in toxic effects, fires, or explosions. The incident impact includes harm to people (injuries, fatalities), harm to the environment, property damage, production losses, and adverse business publicity.*

This disciplined framework is explained in detail in the CCPS Book: *Guidelines for Risk Based Process Safety* (CCPS 2007b). Similar frameworks are available including the Energy Institute's: *High Level Framework for Process Safety Management* (EI 2022). The discipline is generally referred to as Process Safety Management, or PSM, and many companies have their own standards and guidelines for PSM.

An example of a process safety event involving equipment being isolated for maintenance is provided in Case Study 2-1.

Case Study 2-1 — Replacing a Sensor: New Design Leads to a Fire!

Description

An instrument technician was issued a permit to replace a faulty sensor on a turbine batch meter on an LPG pipeline. When the technician undid the screws, the sensor blew out under pressure, resulting in a major loss of containment that led to a jet fire.

Comments

Other batch meters for the process were located outside the pressure envelope and could safely be removed without having to isolate, depressurize, and purge the contents of the pipeline. Neither the operator nor the technician was aware that this batch meter had a different design compared to the others, such that its sensor was located inside the pressure envelope.

Key Learning

- Both the permit issuer and the technician should have been provided with information and training regarding the design of the meter.
- Having two different fundamental designs on the same plant is likely to result in such an event due to human factors (confusion).

2.3.2 Personal Safety

Personal safety (also referred to as Occupational or Workplace Safety) is generally regarded as involving injuries arising from slips, trips, falls, cuts, and bruises, etc. In the context of preparing equipment for maintenance and return to service, it should include prevention efforts for issues such as a fall from height, being struck by an object (such as a pipe that springs when it is disconnected), crushing, trapping of a limb, injury when conducting high pressure tests, or when using pressure jetting equipment. Potential dangers must be clearly communicated to all personnel, as illustrated in Case Study 2-2, involving potential radiation exposure (EPSC 2023).

Case Study 2-2 — No Isolation: Possible Radiation Danger!

Description

A contractor entered a vessel (confined space) and was exposed to radiation from the level detector that had not been isolated.

Comments

The work permit included a question as to whether a radioactive level source required isolation prior to entry, but this was not mentioned on the permit. A standard radiation sign was posted near the vessel entry point.

Key Learning

Permit issuer and receiver must be trained and have the knowledge to understand the hazards and risks associated with the equipment and the work being undertaken.

3 UNDERSTANDING THE HAZARDS

Before the planning process is started, there must be a complete understanding of the hazards to people, the facility, and the environment associated with the materials, process, and equipment that is to be worked on, as well as the task(s) that are to be conducted. This usually requires a team effort since often no one person will have knowledge of all the hazards and activities. For example, process operators would be expected to understand the flammable and toxic nature of the process material and the process' normal operating conditions such as temperatures, pressures, levels, and chemical compositions, but they may not know details of the design or the pressure capability of some test equipment, or whether that machine or tool can potentially produce a spark when it is in operation. Similarly, contractors should have a thorough understanding of their equipment but may not have a thorough appreciation of the risks associated with the process where the equipment is to be used.

Once the hazards are fully understood, the various risk assessments can be performed to identify the appropriate mitigation measures to ensure that the work progresses safely, from the integrity of the initial isolation, decontamination, first line break, and through to testing and recommissioning.

The hazards and required mitigations must be shared/communicated among the workers involved so that they can participate in the Job Safety Analysis (CCOHS online) or other formal risk assessment associated with the scope of work (See Section 4.2.1).

3.1 INFORMATION REQUIREMENTS-PROCESS SAFETY INFORMATION

To understand and evaluate the hazards, the relevant process safety information (PSI) must be available and reviewed by all parties. This information could include, but is not limited to, the following:

- Safety Data Sheets (SDS) for all materials
- Process Flow Diagram (PFD)
- Process Description including safe operating limits (SOL), maximum intended inventories and consequence of deviation (COD) table
- Chemical Reactivity data
- Engineering datasheets showing design pressures, materials of construction
- Relief system design and design basis
- Piping & Instrumentation Diagrams (P&IDs)

- Electrical Area Classification Drawings
- Electrical Drawings
- Isometric drawings
- Details of safety systems and interlocks
- Vessel and equipment drawings
- Relevant scenarios from Process Hazard Analysis (PHA)
- Applicable MOCs since the last PHA

3.2 EQUIPMENT HAZARDS

The process equipment has been designed to operate with specific substances inside it, under a range of temperatures and pressures, which may be expressed as the safe operating envelope. Many facilities specify safe operating limits (SOLs) that are within the safe operating envelope and should not be exceeded without a thorough risk assessment.

Certain operational steps, for example, depressurization of equipment containing hydrocarbons, may lead to excessively low temperatures due to auto-refrigeration that can result in cracking of standard grade steels due to embrittlement.

Some equipment may have a limit set on the maximum rate of temperature change, above which it could become damaged, leading to a possible loss of containment. Examples include:

- Glass lined equipment that requires consideration of the differences in the expansion characteristics of the glass compared to metal.
- Equipment containing packing or catalyst beds where excessive rates of depressurization or pressurization could result in damage to the internals.

Changes in the composition of the fluids within the equipment could result in damage. For example, certain austenitic stainless steel can be sensitive to even very low levels of chloride. Equipment could be damaged due to a material incompatibility with cleaning fluids. Additionally, there is the potential for violent reactions when equipment is opened up, as detailed further in Section 3.3.1.2.

Instrumentation may not be designed for as wide a range of conditions as the process equipment and may fail or be damaged outside its normal operating range. If instrumentation is disconnected in order to conduct decontamination or testing, it must be recommissioned and tested once the work is completed. Level instrumentation requires careful consideration since it is typically calibrated for a particular density of fluid. If there is a change in the fluid, then the level may be significantly different from what is shown on the gauge.

Electricity is present at various voltages up to several kilovolts (kV), and accidental human contact could result in a serious injury, or fatality. The electrical systems and their means of isolation must be thoroughly understood by suitable subject matter experts (SMEs).

There may be covert or hidden stored energy in locations such as behind (or within) isolation valves, non-return valves and other equipment. Other sources of energy include potential (gravitational) energy, such as heavy equipment that requires adequate support, and piping that must be supported when it has been disconnected. Actuated valves and other equipment may be operated by compressed air, hydraulics, springs, or counterweights. Electrical energy may be present, even after isolation where there are capacitors or batteries. Larger plant equipment, such as cooling tower fans might move when the wind picks up, so they should be immobilized and locked.

Relief systems are typically designed to protect equipment from overpressure and must not be isolated in any circumstances without a thorough risk assessment. Again, they must be tested as part of recommissioning.

Further hazards may be introduced by a planned maintenance activity such as disturbance of sediment, mechanical lifting, use of paints, solvents, cleaning materials, etc.

Some items in the plant may contain radioactive level gauges that require removal or isolation.

Temporary equipment, such as rental machines and tools, can create hazards as a result of a lack of hazard recognition that could lead to consequences such as over-pressurizing or overheating. The vacuum rating of equipment must also be understood; many incidents have occurred when vessels have collapsed/imploded when connected to vacuum tankers or drained of liquid at a rate greater than the venting capability of the vessel, or simply due to condensation of steam. Extra consideration is required if the equipment incorporates internal combustion engines, such as diesel/gasoline vehicles, pumps, generators, air compressors, or hydro-blasters, and especially if it is to be used in electrically classified hazardous areas. Jetting equipment can generate electrostatic hazards as well as physical hazards from high pressure fluids.

Use of vacuum tankers for emptying vessels via hoses can lead to the generation of static electricity, providing an ignition source as well as potential injury to workers. Equipment to supply inert gas such as nitrogen for testing brings with it the risk of asphyxiation. Air movers are often brought in to provide ventilation for working inside vessels, which can give rise to electrostatic sparks as well as potentially provide a source of a dangerous contaminant if the location of the air source is not properly considered and monitored.

Other types of equipment hazards include the work location, such as confined spaces, working at height, over water, or near to vent or relief system discharge points.

Further examples of equipment hazards are provided in Table 3-1.

Table 3-1 Equipment Hazard Examples

Equipment Hazard	Justification and/or Examples
Pressure and/or temperature ratings	• Venting a high-pressure material to a pipe or vessel not rated for that pressure could result in a mechanical explosion. • Draining a high temperature material to a plastic tote bin could melt the plastic and cause a loss of primary containment (LOPC). • Cryogenic service adds the risk of cracking, embrittlement, freezing of material and falling ice.
Volume	Transferring materials from a higher volume vessel to a vessel with an unknown volume or one with less volume could result in LOPC due to a vessel overflow or overpressure.
Installed instrumentation	Operators rely on installed instrumentation (pressure, temperature, level, etc.) in the field as well as in the control room, as displayed on the Distributed Control System (DCS), to understand process conditions and to identify potential process hazards. Operations must know which instrumentation is reliable and can be trusted; it is dangerous to rely on unproven pressure gauges for checking the effectiveness of isolations. Sites should implement a process to identify unreliable instrumentation and tag it "out of service" until the instrumentation is either repaired or replaced. If redundant instrumentation is available, it is always best to compare readings to verify process conditions.
Materials of construction	Transferring materials to a vessel with unknown or incompatible material of construction can lead to an acute (immediate) or chronic (over time) failure and could result in a Loss Of Primary Containment (LOPC).

Equipment Hazard	Justification and/or Examples
Equipment & Instrument Bypass	• Each site should have a procedure that defines how the site identifies bypassed equipment and/or instruments. Good practices include, but are not limited to, a bypass board, a bypass book, and documentation of bypassed equipment in shift notes. • Safety Instrumented System (SIS) bypass should follow a rigorous procedure that addresses the risk of bypass.
Out-of-service equipment	Potential to present unique hazards that may include dead-legs or block valves that leak through.
Utility Equipment	• Utility stations should be clearly labeled to avoid inadvertent mixing by using the wrong utility (e.g., using instrument air when nitrogen is intended). • Each type of utility connection should be unique, and a site hose policy should be instituted for each. • Operators should recognize that connecting a low-pressure utility hose to a high-pressure process line could result in hose rupture and loss of primary containment. • Process lines should be de-pressurized prior to utility connection.
Rental equipment	The manufacturer's recommendations and/or equipment owner should be consulted before preparing rental equipment in hazardous service for maintenance or project work.
Contractor supplied equipment	Contractors should be responsible for understanding how to prepare the equipment they supply for maintenance or project work. Risks should be discussed as part of the safe work permitting process and prior to beginning the work authorized by the safe work permit.

Equipment Hazard	Justification and/or Examples
Residual energy in equipment	• Includes, but is not limited to, materials stored under pressure, hot or cold temperatures, electrical hazards, and conditions that could generate fires, explosions, or toxic release. • Operators should identify, understand, and verify equipment hazards resulting from residual energy, such as temperature and pressure ratings of the equipment being maintained or altered and that of any associated equipment being used for decontamination (e.g., hoses, tubing, piping, containers). • Additionally, the accuracy of the instrumentation that is used to determine the presence of residual energy should be verified.

3.2.1 Operational State

Anyone working in a potentially hazardous environment must have a clear understanding of how the hazards may vary based on the state of the process. For example, when the process is "Cold", completely shut down, fully drained, depressurized, purged and tested, such as during a turnaround, the hazards should be relatively low. However, danger can still be lurking in areas such as dead-legs or behind (and sometimes within) closed valves. Also, potential dangers created by auxiliary heating from steam tracing, hot water tracing or electric tracing must be considered. Warm-up procedures may need to include retorquing of flange bolts to avoid leaks (see Section 14.2).

If most of the process is still operational (i.e., "Hot"), and only one section has just been shut down for maintenance, there is a lot of work to do in identifying the risks, planning the isolations, zoning the isolated and decontaminated sections, and specifying the initial opening procedures for the equipment.

The intermediate ("Warm") situation, where a large part of the process is shut down, may require a different degree of planning and activities, although clearly the safest way is to assume that the areas to be worked on are still "Hot".

3.3 ENERGY HAZARDS

A key element of preparing equipment for maintenance and return to service is the safe isolation of sources of energy. Energy is more than just electrical. There are multiple sources of energy associated with the process plant as well as with any equipment that is brought to the site for testing, maintenance, repairs, and recommissioning. The most obvious are the hazards associated with the substances inside the vessels and pipework. These may involve both physical and chemical hazards such as pressure, hot or cold temperature, flammable, toxic or corrosive materials, and potentially unstable/reactive material. Additionally, there are other energy hazards from associated electrical equipment and services, such as steam, electrical power, inert gas for purging or protection of catalyst, water, hydraulics, pneumatics, or spring-loaded devices.

The goal is to understand all of the sources of energy so that measures can be taken to prevent, or at least reduce the risk of injury to personnel, damage to the assets, or harm to the environment. Some of the key energy hazards are provided in Table 3-2; other hazards may be present or even created when performing work associated with the job scope.

Table 3-2 Energy Hazards

Type of Energy	Example of Potential Hazard
Chemical	Injury/ill health/fire/explosion/damage to the environment due to release of hazardous substance (toxic/asphyxiant/flammable/corrosive)
Pressurized liquids and gases	Injury/damage due to sudden release of pressure
Thermal	Injury/damage from hot or cold material
Electrical	Injury/damage due to discharge of energized electrical circuits, static electricity, source of ignition of flammable material
Kinetic	Injury/damage due to movement of rotating equipment or vehicle, crane
Potential	Injury/damage due to fall of equipment from height, release of sprung equipment
Radiation	Injury/environmental damage due to release of radioactive material in the process, radioactive sources, X-rays

3.3.1 Chemical Hazards

Hazardous chemicals can be present as a solid, liquid, and/or vapor. Safety Data Sheets (SDS) must be available for the substances involved and understood by all parties. Safe handling advice is provided within the SDS and many manufacturers offer product stewardship guides for safe handling and storage of the material. Further details on understanding and interpreting SDSs are provided by OSHA 3514 (OSHA 2012).

Chemical Hazards present a risk for personal safety, process safety, and damage to the environment. Process safety incidents, including fires, explosions, and toxic releases can also result in injury, or asset and environmental damage. Again, the measures required to avoid these events focus on keeping the material contained, or at least releasing it in a controlled manner (see Table 3-3 for some hazard examples and potential consequences).

Materials can enter the human body by inhalation (toxic inhalation hazard, (TIH)), ingestion, or absorption through the skin/eyes. The risk of coming into contact with hazardous material is minimized by keeping it contained, but where there is a possibility of a leak, for example during a first line break, it is important to understand the specific type of PPE that should be worn. There are many different types of PPE, and it is essential to specify the correct equipment for the substances involved.

Table 3-3 Chemical Hazard Examples

Chemical Hazard	Justification
Toxic Inhalation	Following an incident involving a LOPC, these chemicals (typically gases) can disperse and affect a large area with consequences including severe injury to many people and may result in adverse impacts to surrounding off-site receptors.
Flammable Gases[1]	Following an incident involving an LOPC, these gases disperse and may result in an explosion or fire. The potential impact area is typically less than the area impacted following an LOPC involving a toxic inhalation hazard but can be devastating to people and equipment located within the potential impact zone.

[1] Including flammable gases stored as liquids under high pressure or cryogenic conditions

Chemical Hazard	Justification
Flammable Liquids	Following an incident involving an LOPC, a release of a flammable liquid can lead to fires and subsequent secondary consequences that may include severe injury and equipment damage.
Other Chemical Hazards	Reactivity hazards: Decontamination may involve draining, venting, or purging of materials that might inadvertently mix with incompatible substances leading to undesirable consequences such as runaway reactions, and fire and explosion. Each site shall have a chemical reactivity spreadsheet, matrix, or table that describes the hazardous effects of mixing of *all* chemicals stored on site. Each decontamination activity associated with potential hazardous mixing scenarios must be understood and plans must be made to avoid inadvertent mixing. Examples: • Purging a flammable with air might lead to a flammable atmosphere and subsequent explosion and/or fire. • Venting Ethylene Oxide to a vent header that contains small amounts of sodium hydroxide might lead to a runaway reaction and explosion. • Draining an acid to a sump that contains a base such as sodium hydroxide could lead to severe reaction. • Draining some acids to a sump and then adding water could lead to a severe reaction. • Some materials such as Rainey nickel catalyst, reformer catalysts, aluminum alkyls and carbon saturated ethylene may burst into flames on exposure to air. • Other chemicals are highly reactive with pure oxygen, hydrogen peroxides, and various organic peroxides. • Removing chemicals from a process by utilizing a vacuum waste truck that may have residual materials within the truck.

Chemical Hazard	Justification
Corrosion	Corrosive Hazards: Purging, venting, and draining chemicals such as corrosives to incompatible material of construction equipment could lead to accelerated corrosion and subsequent loss of primary containment. In some cases, the byproducts of corrosion are flammable.
Embrittlement	Carbon-steel embrittlement: Most light hydrocarbons, such as ethylene, that are stored under pressure undergo a cooling effect when flashed to a lower pressure. If the temperature is low enough (≤ 20 °F) carbon steel can undergo a phase change, whereby it fractures upon re-pressurization.

3.3.1.1 Asphyxiants

Asphyxiants deserve special consideration, since inert gases such as nitrogen may be present during normal operation, or they may be brought in especially for activities such as decontamination, leak testing, or purging before recommissioning. These materials are asphyxiants and even one breath can render a person unconscious. There have been multiple incidents in which a worker has collapsed, sometimes fatally, even after simply looking inside a manway. Often a potential rescuer is also involved in the same tragedy. Further details are provided in the CCPS *Process Safety Beacon* from April 2021 (CCPS 2021).

3.3.1.2 Unstable/ Reactive Materials

Some materials are unstable and may explode upon impact or spontaneously when in contact with air (pyrophoric) or moisture. These materials may be part of the normal process operation, in which case the SDS should provide details of the hazards, or they may be formed as an unintended byproduct. An example of a thermally unstable product that requires special consideration is provided in Section 7.3.

The danger of pyrophoric iron sulfide (PIS) as a source of ignition is a particular issue in refinery and natural gas processing operations. Pyrophoric fires are caused by iron sulfide (FeS) residues, which can ignite spontaneously when air is introduced to the equipment. The sludge/scale can build up on surfaces of vessels, the walls of tanks, or in pipes or tower packing, while other ferrous metal components can contain or be converted to iron sulfide.

This forms during the chemical reaction between hydrogen sulfide (H_2S), which occurs naturally in produced hydrocarbons, and oxidized iron (Fe_2O_3), or rust, that can line a vessel, tank, or pipe. When iron sulfide is exposed to oxygen it reacts rapidly, releasing large amounts of heat.

This exothermic reaction can provide an ignition source for any oil or gas that is present. The oxygen in the air reacts with exposed iron sulfide and eventually reaches a temperature that can cause hydrocarbons to ignite. Other cases have been a slow smolder due to limited air ingress. Of primary importance is preventing iron sulfide from coming into contact with oxygen.

There is also the potential for a PIS-initiated fire in oil and natural gas wastewater tanks. Flammable atmospheres can exist in the head spaces of these tanks due to the presence of dissolved and carried-over hydrocarbons. The recent industry expansion of new sources of oil and gas, involving different production techniques with potentially different impurity profiles, makes such a discussion particularly important at this time.

PIS can also be formed by chemical reaction or by microbial activity. The formation can occur anywhere in the process from the wellhead to the water storage tank. To evaluate risk mitigation options, an operator must first identify the location and mechanism of the formation.

3.3.2 Pressurized Equipment

Vessels and piping may contain fluids under pressure, for example pneumatic or hydraulic valve actuators. Their unintended operation during a period of maintenance might cause a direct injury or affect the integrity of any process isolations. An understanding of how these systems operate and can be safely depressurized or isolated is required.

3.3.3 Thermal Hazards

Hot equipment can cause human tissue damage above about 140 °F (60°C) damage equipment that is brought into contact with it or provide a source of ignition. For example, scaffolding planks can start to smoulder in contact with high pressure steam piping, and electrical or control cabling could melt. Equipment subjected to low temperature can become brittle, crack, or freeze on contact with LNG or other very cold fluids. Similarly, pressurized equipment and piping that contained liquefied gases might reach a very low temperature when depressurized. Gases such as hydrogen have the unusual property of heating up when depressurized (reverse Joule-Thompson effect) and any associated consequences must be considered. A procedure should be in place that specifies heating and cooling of equipment as it is put into or taken out of service.

3.3.4 Electrical Hazards

Electrical hazards from power supplies include direct injury to workers due to burns and shock, or inadvertent operation of equipment. Electrical power can also provide a potential source of ignition in the presence of flammable or combustible materials. Static electricity can be generated in piping and ducting containing high velocity fluids. Electrical energy can remain in systems containing capacitors and batteries, even after isolation. It is important to ensure certain types of equipment are properly grounded. Working on high voltage systems requires the involvement of a subject matter expert and hazards include exposure to electromagnetic radiation.

3.3.5 Kinetic Energy

Hazards from kinetic energy include moving equipment such as rotating equipment, vehicles, and cranes. Equipment that is static could start moving if breaks are released or if affected by wind. For example, a cooling tower fan may be stationary but starts moving when the wind blows. Similarly, an inline ventilation fan's blades may freely rotate when the drive belts are removed from the fan's shaft and pulley. Workers positioned close to operating machinery that is inadequately guarded are at risk of injury.

3.3.6 Potential Energy

Potential energy sources include gravity, compression of a spring or stretching of elastic material. Associated hazards include items dropping from a height, operation of a counterweight (that could open a valve), or a pipe springing when it is disconnected at a flange. The hydraulic head in a tank may create a hazard when a flange or manway is opened at the bottom.

3.3.7 Radiation

Radiological hazards in the oil and allied industries could arise due to the potential presence of Naturally Occurring Radioactive Material (NORM). This may be present in the base of vessels or pipework. Some plants use level gauges that contain radioactive sources, to which a worker might be exposed. X-rays are often used to examine the quality of welds in piping and vessels that can result in workers becoming exposed if they get too close to the source.

Case Study 3.1 — Loss of Containment when Isolating for Maintenance (LyondellBasell La Porte 2021)

Description

On an acetic acid plant, there was a requirement to isolate a section of the process in order to repair part of the pipework. Site personnel chose to use a pneumatically actuated 8-inch plug valve as an isolation device. Company standards required that any valve that is used as an isolation device requires locking in position. In order to lock the valve, the site personnel decided that if they removed its actuator, they could insert a pipe tee over the valve stem and lock it in place, thereby satisfying the company safety requirements, as described in a 2023 report by the U.S. Chemical Safety Board (CSB 2023), and as shown in the diagram (CSB 2023).

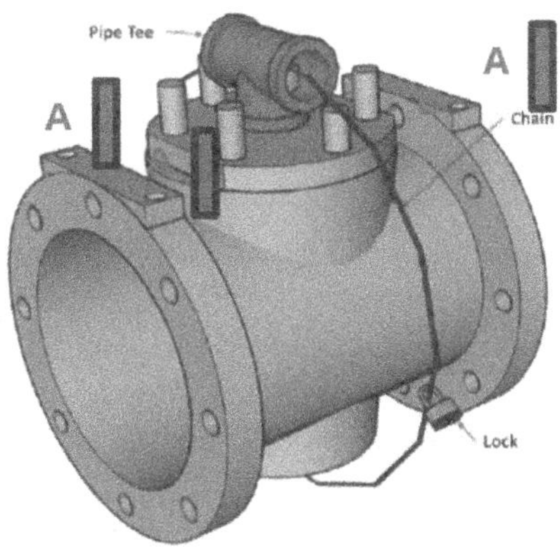

The actuator was bolted onto a mounting bracket and although it was possible to remove it without breaking into the pressurized envelope of the valve, the workers were unaware of this and removed both the actuator mounting bolts (marked A) and the nuts from the pressure retaining bolts (marked B) at the top of the valve. As they were subsequently attempting to remove the coupler on the valve stem, the plug was ejected under the 130 psig pressure from the process fluid, acetic acid. As a consequence, the entire contents of the reactor were ejected, resulting in one fatality, and another worker suffering serious injury.

Comments

CSB reports at least five incidents in which workers have inadvertently broken into the pressure envelope of plug valves because they were unaware of the detailed design of the item installed in the plant.

Key Learning

Workers require detailed knowledge of the design and operation of the equipment that they are working on in order to make safe decisions on work scope. This may require research and involvement of subject matter experts (SMEs).

3.4 SIMULTANEOUS OPERATIONS

During a turnaround, multiple activities take place, and one activity may have an impact on another. This is often referred to as Simultaneous Operations (SIMOPS). This may involve a clash in terms of the location of the activities where, for example, access may be compromised. Multiple work tasks can create conflicting safe work conditions that must be safely managed. For example, hot work and line opening may result in an unexpected consequence. Alternatively, extra demand placed on key services could mean, for example, that nitrogen or water is in short supply and equipment rinsing/purging may take longer than usual. Rental equipment can introduce additional hazards that require consideration, such as diesel fumes entering confined spaces, from welding equipment or emergency power generators.

It may be appropriate at certain times to restrict access, for example, during a first line break, pressure testing, or recommissioning, to essential personnel only.

3.5 HUMAN FACTORS

Turnarounds often involve a huge amount of work and careful planning for such activities. Human factors should be considered during the planning stage, since high workload, complexity, and stress levels can all lead to a greater risk of human error. Nevertheless, there are many measures that can be taken to reduce the possibility of human error throughout this book. Human factors are discussed further in the CCPS book: *Human Factors Handbook for Process Plant Operations* (CCPS 2022a, 2022b).

3.6 FREQUENTLY ENCOUNTERED ISSUES WHEN PREPARING EQUIPMENT FOR ISOLATION

In addition to the items detailed in this chapter, Table 3-4 provides some specific examples of frequently experienced issues that should be understood and considered.

Table 3-4 Frequently Encountered Issues to be Considered

Issue	Details
Obstructions and blockages	Blockage in the line between the block valve and drain. During draining, process fluid could remain trapped. Vents and drains can become blocked and should be carefully rodded once valves are open.
Seized valves	Valves identified in the isolation plan may no longer be operable.
Leaking by/ passing valves	Leaking by/passing valves may not be appropriate for isolation.
Pipe Supports	Existing pipe or equipment supports may not be adequate and additional support may be needed, especially near line break/cutting.
Vacuum effects	Tanks and vessels not designed for vacuum might be sucked in if emptied at a higher rate than design, or if connected to a vacuum tanker.
Environmental Impact	Accidental spillage during draining or flushing can harm the environment.
Chemical Reactions	Cleaning materials might react with a tank or its fittings, leading to potential dangers. For instance, acidic cleaning fluids can corrode certain blinds installed for isolation.
System Overloads	The design capacity of flaring, venting, or draining systems may be exceeded.
Waste Disposal	Safe disposal of pipeline fluids and contaminated water is essential.
Explosions & Fires	Sudden mixing of water with hot oil can lead to explosive reactions.
Explosive Atmospheres	Vapors or gases released during the isolation/first line beak procedure could create an explosive environment. Air can also be introduced into the process during draining.

Issue	Details
Asbestos Handling	Asbestos-containing gaskets should be managed according to safety protocols.
Hydrostatic Load	There is a potential for increased hydrostatic load on pipework and vessels. For example, vessels may be designed for lower density hydrocarbons and might be tested with water.
Ignition Sources	Operations and vehicles in the vicinity can introduce ignition sources.
Asphyxiation Risk	Personnel might be exposed to asphyxiating gases like nitrogen, especially when nitrogen is used to purge the line.
Dead-Leg Hazards	Potential dead-legs in systems can become hazardous regions.
Over-Pressurization & Overfilling	Reception equipment or vessels require protection from excessive pressures or volumes.
Pyrophoric Scales:	In H_2S-containing systems, pyrophoric scale can form and become a fire risk upon exposure to air.
Material Embrittlement	Auto-refrigeration during depressurization can cause valve freezing or embrittlement in steel pipework.

3.7 SUMMARY

Chemical processes pose a variety of hazards as discussed in this chapter that can range from minor to catastrophic consequences. Understanding these hazards and ensuring they are thoroughly addressed in the plans to decontaminate and isolate equipment for maintenance is critical to performing the job task safely. Subject matter experts must be consulted to ensure that all previously known hazards are included in the maintenance preparation plan as well as identifying any additional potential hazards specific to the job task being planned. The equipment decontamination steps should not be initiated until the hazards are fully known, included in the plan, and communicated to affected personnel.

4 PLANNING FOR ISOLATION AND DECONTAMINATION

The isolation and decontamination plan that forms part of the risk assessment must be developed well before the permit to work is issued to the technicians or contractors who will perform the work activity. This will often involve a team effort since a single individual may not have the complete picture of the activities, equipment, and associated risks.

4.1 PLAN CONSIDERATIONS

The plan should clearly identify the scope of the isolation (e.g., a specific vessel, instrument, or piping to be isolated), how this work can best be achieved safely, and the various steps required to achieve it. At the isolation and decontamination stage, it is also important to consider how the de-isolation, leak testing, functional checks on valves and instrumentation and other return to service steps will take place, as there may be opportunities to perform them easier and safer. As stated in Section 2.1, regulatory and environmental aspects must also be considered.

The first stage of the planning process is to ensure that up-to-date process safety information, as detailed in Section 3.1, is available and has been thoroughly checked, so that the hazards of the process are fully understood.

The second stage involves an assessment of the complexity and risks associated with the planned work to determine the degree of rigor required. For relatively straightforward, low risk activities, a local team may be appropriate; for more complex and hazardous work, a multi-disciplined team is required to determine both where the isolations are required in order to conduct the necessary work activity, and how the designated equipment is to be decontaminated and then returned to service once the work is complete. The latter step involves a team of process, maintenance, and electrical personnel, as well as contractors, who will clearly mark up drawings (including P&IDs and/or isometric drawings) and document what should be done as a procedure and/or a checklist. Risk assessments (Section 4.2) must be conducted at this stage to identify and manage the risks associated with those activities. The work area must also be visited ("Walk the Line") to verify the validity of the documentation and the precise locations of the isolations, any purge/drains, line breaks and the work itself, all of which must be clearly identified by a system of tagging or labelling. A summary paper on "Walk the Line" is provided in Appendix B.

All complex and hazardous plans should have validation and documentation of steps in the plan. A key factor to consider when developing the plan is the expected duration of the isolation, which may involve a higher level of isolation and decontamination to be conducted for work activities that have a longer timeframe. Records of changes in isolation plans are essential (duration, scope, accommodating stuck valves, etc.) and each change will require a thorough review of the associated risk assessment.

4.1.1 Isolation Considerations

The planning process must also clearly specify the methods to be used for identifying the isolations, bleed, vent, and line break locations in the field, for example by tagging/labelling/locking. It should also develop/provide the associated documentation that will be required so that all personnel working on the equipment can verify that these isolations are still in place (see Case Study 4-1, which describes the consequences of a line break at the wrong location.) Use of a multi-hasp lock or a lockbox (See Section 8.6) allows all workers to place their personal and/or craft lock on the system, providing a reliable layer of protection where multiple activities are being conducted under a common isolation plan. For longer term isolations, physical disconnection or double block and bleed is highly recommended. Locks and tags are used to complement the isolation devices. Clear identification of isolations is also essential for planning when equipment is being returned to service, to ensure system integrity and correct functioning of instrumentation, including valves, gauges, and interlocks. As part of the planning process, sufficient quantities of suitable isolation and labelling materials must be obtained in advance.

4.1.2 Human Factors

Where the risks are high, human factors and the potential for human error must be considered at all stages, from planning through to the completion of the work. Shift handovers must be thorough, involving clear documentation, discussion and understanding of the scope and timing of the transfer of responsibilities. Activities that are identified as safety critical may require additional scrutiny such as a Safety Critical Task Analysis to ensure a high level of human performance. Further details are available in the CCPS book on Human Factors: *Human Factors Handbook for Process Plant Operations* (CCPS 2022a, 2022b).

4.1.3 Documentation

All isolation/decontamination details must be documented as part of the overall work scope and risk assessment, including highlighted drawings that are clearly marked up to show the stages of isolation and decontamination, as detailed further in Chapter 5.

Case Study 4-1 — Breaking into the Wrong Line

Description

As part of scheduled maintenance, a section of pipeline had to be removed and then replaced. During the planned removal, a contractor cut into a live pipe containing hydrocarbons. The leak was ignited, likely by an angle grinder nearby. The resultant fire overheated an adjacent live ethylene pipeline that ruptured, leading to an explosion that caused four fatalities.

Comments

The contractor had cut into the wrong pipe, instead of cutting the adjacent one that had already been correctly isolated and decontaminated. Three of the four fatalities were firefighters who were caught in the subsequent explosion while fighting the initial blaze, and the fourth was a worker on an adjacent tanker.

Key Learning

As a result of this incident, the site improved its systems for identification and labelling of pipes to reduce the risk of a similar event in the future.

It is not only important to label the isolations, but also the work locations and to make sure that all parties, including contractors, are familiar with the method being used. Human factors come into play when identifying the correct location for work to be carried out.

4.2 RISK ASSESSMENT

Before a safe system of work can be established, there must be an understanding of the potentially hazardous consequences that may arise, and their potential likelihood, from the planned work activities, so that appropriate control measures can be introduced. These control measures may help reduce the probability of an event occurring, the magnitude of the potential consequences, or even entirely change the way the work is carried out. The level and complexity of the risk assessment should be commensurate with the magnitude of the risk. For low risk activities, especially those that have been conducted successfully many times beforehand, a more basic approach is appropriate. Where the risk is higher, or where there is limited experience of the activities, or perhaps where the work spans multiple shifts, a more detailed assessment would be required. As part of the risk assessment, the "hierarchy of controls" must be considered as follows:

- Elimination - physically remove the hazard. This may involve reconsidering the requirement for the work to be conducted.
- Substitution - replace the hazard or use a different approach. In the context of isolation for maintenance, this may involve conducting the work differently, such as carrying out the required work without having to isolate and break into the process. An example might be conducting a pipeline inspection using an external ultrasonic technique.
- Engineering Controls - reduce likelihood or isolate people from the hazard. In addition to relocating personnel to the extent possible, this would be a typical control used for isolation, for example by removing the hazardous material from the equipment, or perhaps reducing its inventory and pressure. Engineering design would feature strongly where, for example, double block and bleed valves are installed at the outset on equipment where maintenance is likely to be required. When the process is being designed and during the initial Process Hazard Analysis, consideration for maintenance isolation and decontamination should be discussed.
- Administrative Controls - change the way people work. This may include reducing the number of people working in the area, extra supervision, or monitoring.
- PPE - protect the worker with equipment. This is particularly important when conducting an initial line/ equipment opening (break-in), where there may be some unexpected hazardous material present, but also during execution of the work where, for example, hazardous deposits may remain inside equipment. During testing and recommissioning, PPE should be worn in case there is an unexpected failure.

Some simple examples of risk control measures are as follows:

- Where there is a risk of personal injury due to a fall when conducting work from a height, the control measure might specify that a harness and fall arrestor should be worn. Alternatively, scaffold access might be specified to reduce the probability of a fall.
- When using rental equipment, additional safety measures may include installing physical barriers, continuous gas monitoring, flame arrestors, insulation of hot surfaces, overspeed trips, and use of standby personnel.

- Where there is the potential for release and accidental ignition of flammable liquid that may be present when opening a valve, one control measure might stipulate the use of non-sparking tools and grounding of equipment. A more analytical process of risk assessment is required where there is a complex network of piping and vessels that require isolation and safe decontamination before maintenance activities can be conducted.

Several tools and techniques that are used to assess risks before work is carried out are listed in this chapter.

4.2.1 Job Safety Analysis

A Job Safety Analysis (JSA) is defined by CCPS (CCPS 2019) as:

> *A procedure that systematically identifies: 1) job steps, 2) specific hazards associated with each job step, and 3) safe job procedures associated with each step to minimize accident potential. Also called job hazard analysis.*

A JSA is sometimes used alongside the work permit and is a more comprehensive risk assessment for higher risk activities. It may be used in association with a written isolation plan and a checklist (CCOHS online). A sample JSA worksheet for changing a flat tire is provided in Table 4-1.

Table 4-1 Example Job Safety Analysis Worksheet

Job Safety Analysis Worksheet		
Job: Replacing a flat tire on a motor vehicle		
Analysis By:	**Reviewed By:**	**Approved By:**
Date:	Date:	Date:
Sequence of Steps	**Potential Incidents or Hazards**	**Preventive Measures**
1. Park vehicle	a) Vehicle too close to passing traffic b) Vehicle on uneven, soft ground c) Vehicle may roll	a) Drive to area well clear of traffic. Turn on emergency flashers. b) Choose a firm, level parking area. c) Apply the parking brake; leave transmission in PARK; place blocks at the front and back of the wheel diagonally opposite to the flat.
2. Remove spare tire and tool kit	a) Strain from lifting spare tire b) Spare tire might be flat or damaged	a) Turn spare into upright position in the wheel well. Using your legs and standing as close as possible to the wheel well, lift the spare out and roll it toward the location of the flat tire.
3. Pry off hub cap and loosen lug bolts (nuts)	a) Hub cap may pop off and hit you b) Lug wrench may slip	a) Pry off the hub cap using steady pressure. b) Use a proper lug wrench; apply steady pressure slowly to loosen and remove the lug bolts (nuts).
And so on...	a) ...	a) ...

4.2.2 Hazard Identification and Risk Analysis (HIRA)/Process Hazards Analysis (PHA)

For higher risk isolations, it may be appropriate to carry out a more formal HIRA (also referred to as PHA). Again, the level and complexity of the study should be commensurate with the risk. These studies usually involve a team approach led/documented by an experienced facilitator and may include the following:

- "What-if" study, where the team responds to standard questions such as: "What if the valve leaks?" or "What if the hose ruptures?"
- Procedural HAZOP study, where the team will examine the procedure and ask questions like: "What if this was done in the wrong order?" or "What if this step was accidentally omitted?" This process can help build more robust procedures that could reduce the possibility of an error (particularly a human error).

Such studies can identify potential weaknesses in, or absence of barrier that would otherwise prevent or mitigate the consequences of adverse events, such as described in Case Study 4-2. The output of the study will include a number of recommendations that should be implemented before the planned isolation is conducted. Further details of these hazard evaluation techniques are detailed in the CCPS Book: *Guidelines for Hazard Evaluation Procedures* (CCPS 2008a).

Case Study 4-2 — Chemical Exposure

Description

When preparing equipment for vessel opening, the work procedure to be used involved utilizing a vacuum-driven elephant trunk system to remove toxic vapors from the equipment. The vacuum would be created by a fan that pulled vapors from the vessel into a neutralization scrubber. The hazards of the process and the health risk of the toxic material were known. The use of the vacuum elephant trunk system had been previously risk evaluated, approved, and safely utilized in the past.

During this accident, the vacuum fan suddenly stopped operating and a backflow of the toxic material exited through the elephant trunk to the atmosphere. A technician monitoring the operation was nearby but was not wearing a protective breathing mask at the time. The work procedure required that the mask should be readily available, but it did not specify that it should be worn continuously. The technician was exposed to the toxic vapors and subsequently died.

Comments

The risk assessment did not include an analysis of the consequences of a vacuum fan failure.

Key Learning

Although a risk review had been conducted, and the work procedure had been safely performed in the past, unexpected conditions were not considered nor discussed with personnel performing the work procedure. A more formal "What-If" type of study may have identified this potential scenario.

4.3 ISOLATION/DECONTAMINATION – WHICH COMES FIRST?

Isolation and decontamination of equipment is not always a linear process. The order and manner of the initial isolation/decontamination must be determined early in the planning process, based on an analysis of the risks, configuration of the equipment, and the practicality of conducting the work. It may be necessary to perform partial isolation and equipment depressurization before the initial line/equipment opening (First Line Break). For example, if only a single isolation valve is available, depressurization of the equipment would be necessary before a blind could be inserted downstream of the valve (See Figure 4-1). This type of operation is typically of short duration and workers may need to wear extra levels of Personal Protective Equipment (PPE).

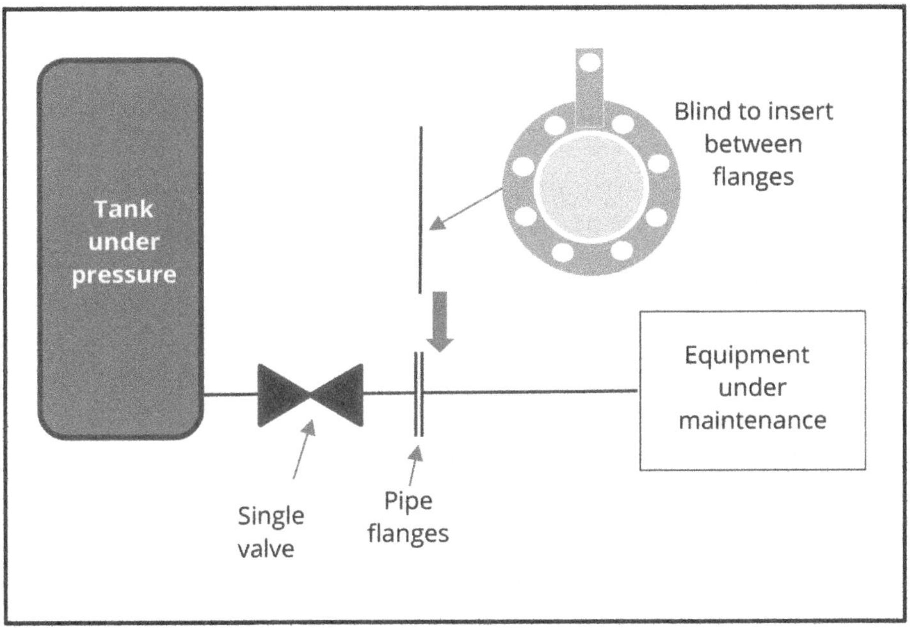

Figure 4-1 First Line Break Against Single Valve

The integrity of the isolation must be proven before additional work can be conducted on the equipment. Alternatively, it may be necessary to remove a plug or cap from a drain or vent valve in order to connect a hose or pipe for draining or purging the process. Often some depressurization/purging/initial decontamination must be conducted before a higher level of isolation (such as removal of a pipe spool) and full decontamination can occur.

4.4 PLANNING THE ENERGY ISOLATION

Many types of isolation are used to close off one part of a system from another, with different levels of ease, effectiveness, and reliability. Some are more appropriate/practical at earlier stages of isolation/decontamination while others are more suited for later, just before work on the equipment commences. The risk assessments conducted at the planning stage determine the level of detail and rigor necessary in the isolation and decontamination plan, including the degree and type of isolation required, and at what stage it will be needed.

Since there may be personal risks when executing the isolation plan, some companies will include triggers or stop gates that are deemed critical to performing the plan safely. A trigger or stop gate may be an action or observation that prompts a worker to stop the work activity and seek help before continuing, in an effort to avoid an injury or environmental event.

Examples of common stop triggers are:

- Discovery of energized equipment not properly isolated,
- Plugged line/equipment,
- Strong odor or discovered leak,
- Excessive force or at-risk body position required to conduct the work,
- Deficient walking/working surface,
- Job task not discussed during the job scope or person inexperienced at job step identified,
- Job scope changes, or hurrying/rushing encountered.
- Poor visibility or lighting.
- Lack of egress.

These types of stop triggers should be discussed with personnel before beginning the job and informing the team that they have the authority to stop work if they observe an unsafe condition.

Typical examples of *process line isolations* include single valve, double valve, double block and bleed blinds, or physical disconnection, as detailed further in Section 10.1. It is important to include isolation of services lines including nitrogen, steam, air, water, and process drains. If suitable valves and blind-points are not available, special techniques for process isolations as detailed in Section 10.6 can be used, but only under very carefully managed conditions, where more reliable options are not available. Additionally, changes in isolation plans can result in serious outcomes if only communicated informally, as described in Case Study 4-3.

Examples of *electrical isolation* include disconnecting the power supply via a switch, removing fuses, or physically disconnecting power cords. Once the isolation step is completed, the system should be proven to be dead/inactive. For high voltage systems, subject matter experts (SMEs) must be involved. For lower voltage systems, it may be appropriate to verify the isolation (prove it dead) by trying to start the equipment in a safe manner, which is sometimes referred to as Lock-Out Tag-Out Try-Out. Further details are provided in Section 10.2.

Examples of *instrumentation isolations* include disconnecting energy sources (air/nitrogen/electrical) to automatic shutdown valves and control valves or other machinery actuators. Instrument isolations may require bypasses/inhibits, etc. that require a formal management system of checks and balances and additional temporary measures. These are further discussed in Section 10.1.

It may also be appropriate to physically lock equipment that could otherwise move, such as fin-fans that rotate when the wind blows, springs, or counterweights, as discussed in Section 10.3.

Case Study 4-3 — Isolation on an Oil Refinery

Description

Operations personnel were preparing equipment for maintenance on an oil refinery (CSB 2017c.) A single block valve had been closed between the work location and the Depropanizer column. One of the workers began draining what was assumed to be water into the oil water sewer when a flash fire erupted, causing him serious burns.

Comments

The single isolation valve was leaking by (gas passing through the valve), allowing hydrocarbons from the Depropanizer column to flow through to the drain point. A nearby furnace ignited the gas.

Due to problems with another passing valve, the facility had changed its isolation plan from a relatively routine one, to a much larger, more complex one. This plan was developed informally and was not subject to a detailed risk assessment.

Key Learning(s)

- Sites should have general procedures for preparing equipment for maintenance and processes for identifying and addressing potential hazards associated with those tasks.
- Procedures should be clearly communicated to all who will be responsible for performing the work.
- Companies should avoid using a single block valve and consider more protective methods of isolation when preparing equipment for maintenance.
- A management of change system should have been used when the scope of the isolation changed. This would have included a rigorous risk assessment that should have identified the associated hazards that required proper management.

4.5 PLANNING FOR DECONTAMINATION AND DECOMMISSIONING

Decontamination may be carried out for a specific maintenance activity and is also required for decommissioning, where a plant (or part of a plant) is being taken out of service, potentially indefinitely. While much of the initial work may be similar, additional considerations will be required in the planning stages for decommissioning that should include a risk assessment commensurate with risk. A Management of Change (MOC) process would be the minimum requirement for minor decommissioning and a more detailed methodology such as HAZOP/LOPA would be appropriate for more complex decommissioning. Different standards may apply where equipment is being decommissioned permanently in advance of demolition, versus "mothballing" where it might be used one day in the future. The latter case may require a formal preservation philosophy to be developed to include instrumentation, static and rotating equipment. This would involve additional consideration, such as filling the equipment with an inert gas, and maintaining a slight positive pressure, to preserve its condition in the longer term. (Case Study 4-4, Case Study 4-5, and Case Study 4-6 describe the real-life consequences of improper long term isolation /mothballing.)

Case Study 4-4 — Isolated but Not Fully Decommissioned - 1

Description

Auditors discovered a pressure vessel still contained a "heel" of pressurized flammable liquid. Although the vessel was isolated from the process, it had not been decontaminated. Since the vessel was no longer used, the facility no longer included the safety relief valve in its preventive maintenance test schedule. Further discussions identified that this had been the case for several years and was considered a low priority.

Comments

The vessel was effectively still in service although it was isolated from the process. It was potentially not adequately protected since its relief device had not been tested. As part of the follow-up recommendations, the auditors designated fully decontaminating the vessel as a high priority action item.

Key Learning

Equipment that is no longer used must be formally decommissioned via a managed project or a management of change procedure. This should include vessels as well as piping dead-legs that are known causes of major incidents in the process industries.

Case Study 4-5 — Left a Valve Dead-leg - Not Fully Decommissioned - 2

Description

A major loss of containment of LPG and fire occurred at Valero's McKee refinery in February 2007 (CSB 2008), leading to injuries to four personnel and major damage to the facility. The incident occurred due to water freezing in a section of 10-inch pipe that was isolated and had been out of service for about fifteen years.

Comments

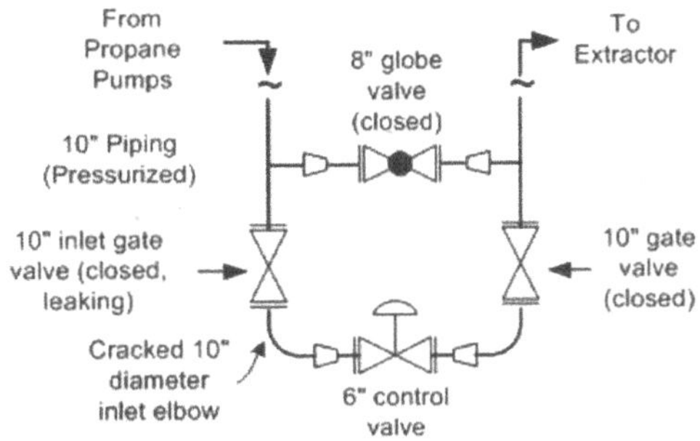

A change in the configuration of the process meant that a section of piping including a control valve and associated isolation valves were no longer in service. No MOC review was conducted. The section had been isolated via manual isolation valves forming a "dead-leg", although unbeknown to the operators, one of the valves was leaking by. During a particularly cold spell, water that was normally present in the process and tended to accumulate at low-points, froze on a bend downstream of the manual valve. This caused the elbow to rupture and once the weather warmed, the ice-plug melted, causing the loss of containment of liquid propane, which ignited.

Key Learning(s)

- A full MOC must be carried out when there is a change in process operations.
- Where dead-legs arise, they should be physically isolated by blinds (as a minimum) and ideally physically disconnected.

Case Study 4-6 — Isolated Longer than Intended

Description

A major loss of containment of butadiene occurred at TPC's facility in Port Neches in November 2019 (CSB 2019). The butadiene subsequently ignited, damaging further equipment and leading to a series of subsequent explosions. There were two injuries, and the butadiene unit was permanently closed.

Comments

The incident occurred due to "popcorn polymer" forming in a section of piping associated with a pump that had been taken out of service for repairs that took longer than expected - 114 days until the day of the failure. Formation of "popcorn polymer" is a known phenomenon in areas of low or zero flow and is particularly dangerous since it expands and can lead to pipe rupture, as occurred here. The pump repairs took longer than expected allowing the polymer to form and accumulate in what had effectively become a temporary "dead-leg". A HAZOP that was conducted three years prior to the incident had recommended flushing of temporary out-of-service equipment on a monthly basis, but the recommendations had not been implemented.

Key Learning(s)

- When equipment is temporarily isolated for maintenance, but not fully decontaminated, any potential risks must be identified, particularly when "dead-legs" are formed.
- If the duration of isolation changes beyond what was originally expected, further risk analysis must be conducted.
- HAZOP recommendations must be implemented in a timely manner.

4.6 DISPOSAL OF CHEMICALS FROM DECONTAMINATED PROCESSES

As part of the planning phase for decontamination, it is crucial to determine where removed materials can be safely disposed of or routed. This is of special concern if hazardous materials must be taken off-site, and additional disposal steps are necessary. Off-site disposal will require coordination planning with various on-site personnel and off-site organizations, such as environmental department, scheduling, shipping, local, and regulatory authorities.

Off-site support and considerations must be provided for arranging for any specialty equipment that may be required for decontamination. Examples could include hydroblast systems for cleaning piping, portable steam generators, nitrogen tanks, vacuum trucks for waste management, chemical holding tanks for temporary storage, or chemical circulators for treatment processes. It is often preferrable to arrange for and manage disposal with equipment designed and approved for specific chemicals, but if off-site coordination is needed because the facility lacks either the necessary equipment or the capacity to manage the volume of decontamination material, the decontamination plan must ensure safe and environmentally sound off-site disposal.

4.7 PLANNING FOR RETURN TO SERVICE

The isolation/decontamination planning must also include the activities to be conducted after the maintenance work is completed. Associated procedures can range from routine/normal return to service, a temporary procedure, standard operating procedure (SOP), or checklist SOP with independent verification. Again, the level of planning and risk assessment has to be commensurate with the hazards of the process and the complexity of the activities involved that include inspection, purging, testing, drying, and other recommissioning processes. It is also important to establish manning requirements, while limiting the presence of non-essential personnel, and to identify additional equipment that may be required such as vacuum trucks for spills, additional environmental monitoring for possible leaks. Further details are provided in Chapter 11 - 13 and summarized in Table 4-2.

Table 4-2 Planning for Return to Service

Step	Details of Activities
Preliminary Verification	Pre-start Safety Review: Conduct a comprehensive review to confirm all maintenance, repairs, or adjustments have been properly completed. Validate that no tools, debris, or personnel are inside or near the equipment that could cause issues during startup. Maintenance Checks: Equipment Safety Verification: Maintenance team confirms that the equipment is mechanically sound and safe to start. Example 1: Energize pumps to verify rotation direction is correct. Example 2: Conduct functional tests of sensors and safety mechanisms to ensure they are working correctly.
Isolation Removal	Removal of Locks and Tags: Ensure that all personnel are cleared from the immediate area and confirm that it is safe to remove lockout/tagout devices. Follow organizational procedures for removal, documenting each step.
De-isolation of Equipment	Remove all isolation measures like blinds, or other barriers used for process isolation. This should be done in accordance with established procedures and should be double-checked.
Pre-Start Activities	Air Freeing/Purging: Remove any trapped air or inert gas from the system as required. Ensure the process lines are clear.
Introduction of Process Fluids	Slowly introduce process fluids or media to the system, observing for any leaks or issues that may arise.
Equipment Startup	Initial Startup: Energize the equipment at a low operating level initially, monitoring all systems for abnormal conditions.

Step	Details of Activities
Operational Tests	Service Testing: Carry out operational tests to verify that the equipment meets the service requirements. This may include functionality, efficiency, and safety checks. Temperature/Pressure Stabilization: Gradually run the equipment up to its operational temperature, monitoring all relevant parameters for stability and safety.

The plans should consider issues that can arise during removal of isolation. For example, if the original isolation was achieved using a single valve followed by a blind, there is the potential for the space between the valve and the blind to become pressurized, if the valve was passing leaking by slightly. This problem could be reduced by removing a pipe spool, and then fitting a blind flange incorporating a small bleed/purge valve onto the flange (that should be fitted with a plug and suitable label), see Figure 4-2.

The planning process must consider this type of potential issue and ensure suitable plans are in place to allow safe de-isolation.

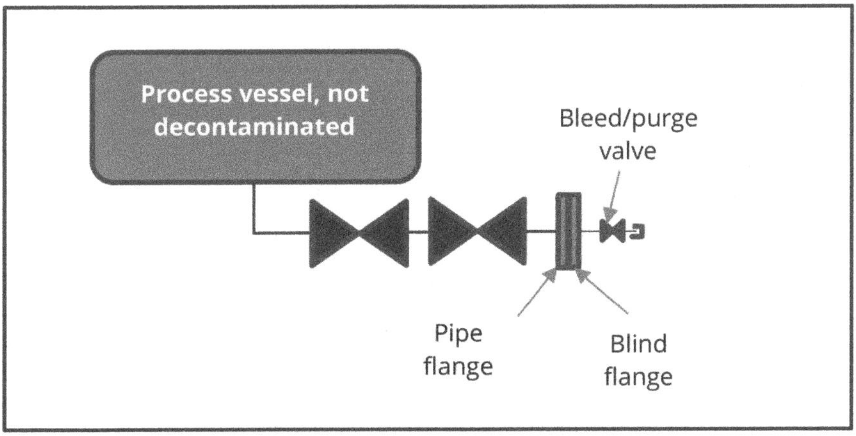

Figure 4-2 Purge Valve Fitted to Blind Flange

4.8 CHANGING PLANS

If there are any changes to the plans, it is essential to review and re-address with the team the risk assessments associated with isolation, decontamination, testing, and recommissioning. This can be conducted using a Management of Change (MOC) procedure, or a temporary MOC.

Examples of such changes include the following:

- Planned isolation cannot be achieved due to faulty equipment (e.g., leaking by/passing valve). It may be necessary to shut down a larger section of the plant to achieve safe isolation.
- Maintenance work scope is extended or changed to additional equipment/pipework.
- Duration of work is extended beyond the original plan.
- Decontamination process changes, e.g., change of nitrogen supply from plan nitrogen to bottled.
- Testing plan changes due to availability of equipment.
- Recommissioning process changes due to lack of availability of materials or equipment.

5 THE DECONTAMINATION PLAN

5.1 DOCUMENTED PLAN

A documented plan must be developed to ensure that the equipment does not present a hazard when it is opened up to the atmosphere. The level of detail within the plan and its approval requirements must be commensurate with the complexity of the decontamination and associated hazards. Some key components of the plans are shown in Table 5-1. The plan and associated procedures require careful preparation and the involvement of plant personnel and any contractors who will be responsible for carrying out the work. These procedures must be written down and must include the supporting documents and drawings that will form part of the line/equipment opening permit. With the exception of very simple, low risk activities, the use of clearly marked up P&IDs and/or isometric drawings is essential to ensure personnel fully understand the planned activities. Under some circumstances, e.g., relatively straightforward decontamination operations such as connecting drain hoses, purge tubing, or pipe for flushing, use of temporary Management of Change procedures may be appropriate, but they, too, must be carefully documented and clearly communicated to the workers.

The procedures and associated permit must specify the PPE required, placement of temporary barriers, any appropriate emergency procedures or provision of standby emergency response crews and equipment in case of unforeseen problems, as well as any special warnings or provisions associated with opening the equipment. The use of checklists is recommended at all stages of decontamination and should form part of an overall checklist up to returning the equipment to service.

Plans must be reviewed and signed off by trained, qualified and competent staff. Approval at a high level may be required if standards of decontamination do not reach company standards or best practice. For example, if the workers cannot verify that a system has been fully decontaminated of a toxic substance prior to initial break-in, a senior manager, possibly in association with a relevant subject matter expert (SME), may need to be called in to discuss options, and to review and approve the mitigating measures that are proposed (see Case Study 5-1, where additional risk assessment by senior management was required for a first line break.)

In some situations, routine maintenance on an item is required but decontamination is not possible due to design or process constraints—for example, removing a conservation vent on an atmospheric storage tank for preventive maintenance or replacing a faulty instrument in a chemical sump. Under such circumstances, a documented, approved work practice to remove each device should be prepared in advance and used when these specific activities take place.

Table 5-1 Key Components of Decontamination Plan

Consideration	Issues
Hazards of the process	Chemical and other energy hazards including toxicity, flammability, unstable materials, trapped energy
Risk Assessment	Assess risk commensurate with the hazards and complexity of decontamination.
Written Plan	Based on risk, determine how the plan is documented and communicated to affected personnel. Decontamination plans that are executed across multiple units and shifts require more formal documentation, communication, and tracking. Determine where and how progress against plan is documented and communicated.
Purge medium	Optimal purge material, purge rate, pressure/ depressurize, dead spots, testing
Hoses, vents and drains	Materials of construction of hoses, avoiding mixing up hoses (color, fitting type).
Vents and drains	Use of first line break procedure when opening. Clear definition and identification, vent and drain line materials/interactions, capacities.
Tool requirements	Joints, fittings, pressure gauges (calibrated)
Material hold-up	Potential for residual material, e.g., solid/liquid deposits, trapped substances, material held on catalyst/adsorbents.
Testing	Analysis to determine effectiveness of decontamination, retesting/bleeding at appropriate intervals to confirm decontamination/check for valves that are leaking by/passing.
Line break	First line break procedures - out of the line of fire, PPE, essential personnel only

5.2 PHASE OF DECONTAMINATION

It may be appropriate to perform the decontamination procedure in various phases, as the degree or integrity of the isolation from the process progresses. When completing a first line break, for example, it may be necessary to first insert a blind or a blind flange in a line downstream of one or more closed valves, where there is either no or limited opportunity to decontaminate a small part of the pipework. This is discussed further in Section 4.2 and should be avoided if possible as it would require special provisions, as outlined in Section 5.1.

Case Study 5-1 — Toxic Break-in

Description

A chlorine system was designed with double-block valves (A and D) and a bleed valve (C) fitted in between, complete with a pressure gauge, so it was possible to test the integrity of the first isolation valve. Normally, use of the double block and bleed would have satisfied the corporate requirements for working on the downstream equipment and a permit issued by the plant superintendent would have been appropriate. However, it was established, first from a pressure reading on the gauge (PI) attached to valve (B) and then by cracking open the adjacent bleed valve (C), that the main valve (A) was not providing an adequate seal. It was not possible to empty the upstream vessel of chlorine without using the equipment downstream that required maintenance, so a break-in against a single isolation was required, prior to insertion of a blind. The company procedures required the risk assessment to be escalated to more senior levels of management.

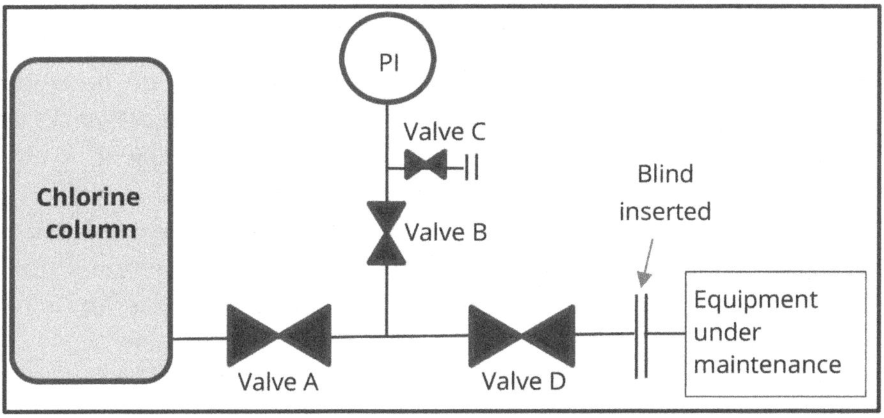

Comments

The risk assessment required a number of mitigating measures that included:

- Decompression of the Chlorine column as much as possible.
- Valve C connected to the scrubber line to reduce pressure between A and D as much as possible.
- Conducting the break-in with fitter wearing a full air-fed suit.
- Standby emergency response team with air-fed suits and water mitigation sprays.
- Work conducted at night when only essential personnel were present.
- Wind direction was out to sea, away from the populated area.

Once the maintenance work on the downstream equipment was complete, similar precautions were required for the blind removal. It then was possible to fully empty the chlorine column and evacuate it before valve (A) was replaced with a new valve that could provide a reliable seal.

Key Learning(s)

- When corporate isolation standards cannot be met, additional risk assessments must be conducted at higher levels of management and/or using subject matter experts.
- An alternative solution may involve shutting down and decontaminating a larger part of the facility.

5.3 MANAGING THE DECONTAMINATION HAZARDS AND RISKS

The decontamination process is aimed at removing as much of the hazardous material as possible to reduce the risk of exposure to personnel, the environment or incompatible equipment. Where it is not possible to remove hazardous material(s) entirely, there should be an understanding of what could still be present, and the risks and mitigations required to prevent injury to people, property, or the environment. Further details on understanding the hazards are provided in Chapter 3, and further, specific areas to consider are as follows:

- Break-ins (First line breaks) must be clearly identified and labelled. Only have essential staff in the area with appropriate PPE.
- Use checklists and/or clear records, to ensure that the status of decontamination is communicated and understood between successive shift teams.
- Do not rely on readings from local pressure gauges unless they have just been properly calibrated.
- Clearly define and document the depressurizing confirmation plan.
- Carefully consider pressure relief systems, especially where they may have been isolated, or where purging systems can potentially over-pressurize equipment and piping.
- Beware of non-return valves and other equipment that could hold material within or behind them.
- Be aware that hazardous material may still be present even after depressurizing and initial purging (see Case Study 5-2 for an example of a tank that exploded before being properly decontaminated for demolition) (CCPS 2008b)). This is especially important where there are unstable/pyrophoric materials present (an example is described in Case Study 5-5).
- Where lines or valves are plugged (blocked), process and equipment specialists' advice should be sought, and a further risk assessment should be undertaken.
- Consider Simultaneous Operations (SIMOPS) when purging to a vent system. If another decontamination activity is being conducted, there may be a sudden reverse flow from the vent system. Check for potential incompatible materials (chemical reactivity) where different substances are mixing in vent headers.
- Take care when using hoses. They must be in good condition, recently inspected for integrity and preferably color-coded so there can be no mix-up between nitrogen, air, water, etc. It is recommended to use hose adaptors that are specific to each utility.
- Where hoses might operate under pressure, they should be fitted with whip-checks (to prevent injury if the joint fails) and designed to handle the maximum pressure of the delivery system or the process.
- Consider materials of construction issues when flushing and decontaminating equipment. For example, certain austenitic stainless steels may be susceptible to chloride induced stress corrosion cracking (CISCC) if put back into service with even very low levels of chloride on the surface that may remain from the decontamination operation.

Case Study 5-2 — Refinery Storage Tank Explosion

Description

A floating roof petroleum storage tank exploded as it was being prepared for demolition, resulting in one fatality and serious injury to a second worker. The incident occurred when workers were attempting to remove a breather valve from the roof.

Comments

The tank had not contained petroleum products for some time, having been used as a process water surge tank to support the refinery waste treatment system. However, it likely contained some flammable vapors, possibly hydrogen due to internal corrosion. Removal of the breather valve was initially attempted by mechanical means, which the site considered to be a safe method, but that was not effective. A contractor was then called upon to use a portable electric saw. The incident report does not indicate if a hot work permit had been prepared and issued.

Key Learning(s)

- There was a failure to understand that possible hazards such as flammable vapors could be inside the tank. Something as routine as using a spanner (wrench), especially if hit by a hammer, could have generated a spark.
- Use of a hot work permit and suitable flammable gas test would have identified the hazard, assuming an analysis was made of the contents *inside* the tank. Similar incidents have occurred where only external gas tests were conducted.
- The work plan for removing the breather valve did not consider an alternative means to safely remove the breather valve if unbolting was not successful.
- A change in the work plan should have initiated a revised risk assessment.
- A written isolation and decontamination plan might have flagged the risks and prevented this incident.

5.4 TRAINING FOR DECONTAMINATION

Specific training that includes methods of decontamination (steaming, purging, inerting, using dry air, etc.) may be required. Contractors must be included in the training, particularly where they are involved in an initial break-in (first line break) or use of hoses and couplings.

An example of a more challenging decontamination method is provided in Case Study 5-3. A complete lack of a decontamination plan is shown in Case Study 5-4.

Case Study 5-3 — Solids Plugged in a Thermosyphon Reboiler

Description

A thermosyphon reboiler installed on a distillation column that processed toxic materials was subject to plugging with solids during normal operation. Several tubes would plug solid, but not all of them. This made preparing the reboiler for cleaning a very challenging procedure.

Comment

The reboiler required double block and bleed valves on both the inlet and the outlet sides. The cleaning process began with depressurizing the column and reboiler; then isolating the reboiler with the block valves and draining some of the liquid to a safe collection tank. Residual liquid and vapors remained in the reboiler. A special decontamination procedure was then required for further cleaning, which involved purging with nitrogen, re-heating the reboiler, and monitoring for contaminants in the purge gas. Once the contaminants in the purge gas reached a level considered safe for opening the reboiler, senior leadership would approve the equipment opening permit. Although this procedure was carefully developed and the associated risks previously considered, there remained potential for exposure consequences that were re-reviewed with personnel each time the reboiler was to be opened.

Key Learning(s)

- Equipment opening plan must be considered, developed, and reviewed on a case-by-case basis.
- Standard practices for decontaminating equipment must consider the presence of residual materials, including solids.

- Personnel must be made aware of the hazards and special precautions necessary to prevent exposures.
- Sometimes, equipment cannot be fully de-contaminated before opening and additional precautions must be included in the permit.

Case Study 5-4 — The Tank That Was Not Isolated

Description

A low-pressure intermediate process tank contained a mix of hydrocarbon materials and was connected to a common vent header that was directed to a vapor recovery system. The vent header had a manually set nitrogen purge flow meter located at the end of the header. An explosion in the tank occurred after the tank had been emptied in order to replace a faulty level probe located on the bottom side wall of the tank.

Comments

This event occurred during a plantwide turnaround. Although it had been emptied and drained, the tank was not purged free of vapors. There were no plans to open the tank or to enter the tank. The job scope was to simply unbolt the level device from a bottom tank nozzle and install an identical replacement. The level device had been removed, but replacement was delayed. The explosion occurred when the level device nozzle was open and unattended. Since no personnel were present, serious harm was avoided.

The incident investigation found that there was no isolation or decontamination plan documented for the tank, there was no written maintenance procedure for replacing the level device, and the nitrogen purge system on the vent header was plugged, indicating there was no nitrogen blanket on the tank. A further key factor for the explosion was that several years earlier, the flow into this intermediate process tank had been modified, resulting in an increase of hydrogen in the liquid stream. An MOC had been conducted for the process flow change, but the MOC failed to recognize the increased H_2 concentration. The source of ignition was not identified, although it may have been due to the static ignition of the hydrogen in the head space of the tank.

Key Learning

- Responsible personnel must understand the risks and create and review an isolation-decontamination plan.
- A line opening permit must be used for the replacement of the level device, which would have prompted questions about the status of the tank.
- MOCs approving process flow changes must consider the composition change of the materials introduced to the tank.
- Preventive Maintenance procedures should be established for routine tasks such as the inspection of the nitrogen purge system on the vent header.

5.5 DIFFICULT DECONTAMINATIONS

Extra consideration may be required for processes that tend to plug, or that use solids or viscous liquids that can prevent or restrict safe draining or venting of the process and equipment. Liquids that are normally pressurized, but vaporize at atmospheric pressure, can auto-refrigerate (Joule-Thompson effect) and form liquid pools even when depressurized. Such systems can be particularly challenging to decontaminate. Sometimes decontamination can be achieved by successive pressurization and depressurization, for example, where the equipment contains catalysts or adsorbents.

Remaining materials may be toxic, flammable, thermally unstable, or possibly pyrophoric. The hazard of pyrophoric iron sulfide as a source of ignition in refineries and natural gas wastewater tanks is discussed further in Section 3.3.1.2. Decontamination techniques include chemical and mechanical methods. One example is provided in Case Study 5-5 (Ender 2003).

One group of particularly challenging compounds is alkyl lithium. These materials are toxic, pyrophoric, and typically present in a flammable carrier solvent. They react with moisture, emitting hydrogen and decompose in air, releasing heat. If allowed to settle, solid material is formed that can be very difficult to remove. An example decontamination method that uses "reactive cleaning", where controlled amounts of moisture are added to promote a reaction, is provided in Chapter 7.

Case Study 5-5 — Pyrophoric Ignition of Structured Packing in a Crude Vacuum Column

Description

A refinery was shut down for maintenance, including work on the crude vacuum distillation column. After two days of steaming out and then cooling the column, the manways were opened, starting from the bottom. After opening the manway second from the top, workers noticed smoke coming out of the manway and glowing inside the structured steel packing. They closed the manways and resumed steaming and then cooled the column with water. The next day, they opened up the column again and saw that one of the packed beds had been destroyed by fire.

Comments

The fire was caused by iron sulfide deposits on the packing that self-ignited when exposed to air. The refinery had switched to a higher sulfide crude about a year previously.

Lessons Learned

- Where there is a risk of pyrophoric material remaining in equipment, steaming out may not be sufficient for full decontamination.
- Carry out a risk assessment and establish a rigorous decontamination procedure that may include chemical cleaning. In the case of pyrophoric iron sulfide (PIS), this may include permanganate or percarbonate wash, purging, and adequate cooling.
- Open the column carefully, minimizing the airways and monitor temperature closely. Document emergency procedures, to include standby fire crews and a process for rapid reclosure of the column if there is any sign of a temperature rise.
- Carry out a full management of change (MOC) process when raw materials change. The impact of a change can often be hidden for years, where there is a change in the chemistry of the process.

5.6 MOVING DECONTAMINATED EQUIPMENT

Entire plants and associated machinery, such as pumps, pipework, valves, pressure relief devices (PRDs), and compressors may require removal from the immediate work area and although such equipment may have been decontaminated sufficiently for removal, it could still contain quantities of hazardous material/energy. For example, material could be trapped in pockets, beneath a liner or within a valve. Sometimes this does not become apparent until the equipment is opened up, which may take place in a workshop or even offsite at a contractor's premises.

The site may have facilities for performing additional decontamination, for example, by steaming the equipment out in a decontamination bay. However, since this extra step still may not guarantee that the equipment is completely free from hazards, this must be formally communicated to those carrying out the work at some stage in the future. Typically, this would involve providing a Decontamination Certificate that lists substances that may still remain inside the equipment. An example of a Decontamination Certificate is provided in Appendix C. Locations where additional decontamination is carried out remote from the plant require an associated material waste/disposal plan as further discussed in Section 4.6.

5.7 CHECKLISTS AND EQUIPMENT IDENTIFICATION

Checklists and independent verification may be appropriate, depending on the level of risk to the worker and the environment, especially for complex decontamination plans across multiple units and shifts. A system of positive equipment identification (PEI) is required, including "Walking the Line" in order to ensure the validity of the documentation and the correct work location. The initial line break should be carefully labelled, and detailed records must be kept regarding all isolations, the position of vent/bleed valves, as well as purge points and connections. This will help to ensure all parts of the system are restored to the correct position and that the temporary purge and vent lines are removed and suitably isolated when the equipment is placed into service afterward.

6 EXECUTING THE DECONTAMINATION PLAN

6.1 SHARING THE PLAN

The decontamination plan must be shared with the multiple parties/ stakeholders involved so they are all aware of the key activities required both before and during the decontamination process. The extent of the involvement of other groups and individuals should be commensurate with the risk. For higher risk decontaminations, the parties who should be fully aware of the decontamination plan include the following:

- Operations team
- Maintenance team
- Instrument personnel
- Electrical personnel
- Contractors
- Nearby personnel who may be affected
- Connecting and adjacent process facilities if affected
- Emergency Personnel
- Safety, health, and environmental personnel

After the decontamination is completed, it may be necessary to carry out further monitoring as the work progresses. It is essential that everyone is aware of who carries out this monitoring, how it is accomplished, and at what frequency. This is especially important when entry is to be made into confined spaces such as tanks and vessels, when re-testing should be conducted at least daily and more frequently, if for example, deposits are present that may release dangerous substances if disturbed. Even the most basic decontamination plan may affect other activities that are in progress or are planned at the facility. This is more likely if the facility is beginning a maintenance turnaround with several concurrent planned maintenance projects (SIMOPS). In this case, a thorough understanding of multiple decontamination plans must be shared and coordinated via a rigid process that is accessible to plant personnel and prominently posted in the facility. As a minimum, permits must provide contact personnel and cross-references between activities in the same area or on the same system.

6.2 TRACKING THE PLAN

The decontamination plan requires tracking to ensure that every stage listed in the plan is completed. A checklist, (see example shown in Table 6-1 at the end of this chapter) is essential for more complex or higher risk situations, and dual sign-off may be required for more critical stages. Verification of decontamination including for example, the measurement of gas levels, should be conducted as appropriate to the risks presented by the substances involved. Any changes to the original plan must be fully documented, showing any additional risk assessments that have been conducted as a result of the change. Because planned activities may bridge across multiple operating shifts, the facility must have a process for clearly communicating the current status of all activities, issues encountered, and any changes that have been made to the plan during the previous shift.

Coordination of multiple decontamination plans will also pose special considerations for tracking. Since confusion can occur as to the status of the individual plans, maintaining and documenting an individual checklist for each plan is essential to completing all decontamination steps in the approved order.

Once the equipment has been successfully decontaminated, isolation barriers must be verified to prevent re-contamination, which may include pipe caps and other devices designed to keep foreign material out of the equipment. A detailed list/description of barriers and verification of their status must be included in the decontamination plan. This is especially critical during Transient Operations such as facility shutdowns, when multiple activities are occurring simultaneously, as described in Case Study 6-1 and shown in Figure 6-1 (CSB 2022). Further information on Transient Operations can be found in the CCPS book, *Process Safety During the Transient Operating Mode* (CCPS 2021). Details of the isolation barriers must also be incorporated into the recommissioning plan to ensure they are removed before the equipment is returned to service.

Case Study 6-1 — Explosion/Fires at Refinery During Maintenance Turnaround

(Husky Superior in Wisconsin April 2018)

Description

On April 26, 2018, an explosion occurred at Husky Energy's Superior Refining Company LLC refinery in Superior, Wisconsin ("Husky Superior Refinery"), resulting in subsequent fires and injuries to 36 refinery and contract workers. An estimated 39,000 pounds of flammable hydrocarbon vapor mixture was released during the event. The incident occurred during a planned maintenance event (turnaround), while shutting down the fluid catalytic cracking ("FCC") unit. Two FCC unit vessels—the primary absorber and the sponge absorber—exploded, shaking buildings up to a mile away and propelling over 100 metal fragments, some several feet long, up to 1,200 feet from their original location, into the surrounding operating areas.

In addition, explosion debris punctured a nearby asphalt tank at the refinery, spilling hot asphalt that flowed outside of the tank's containment area. Approximately 17,000 barrels of hot asphalt spread through the refinery and ignited, causing fires to erupt at multiple operating areas of the refinery. The City of Superior evacuated 2,507 residents within 2 miles north, 3 miles to the east and west, and 10 miles south of the refinery.

Figure 6-1 Fire at Husky Refinery (CSB, 2022)

Comments

The incident occurred while the FCC unit was being shut down, which is a period of transient operations that presents additional hazards not normally present during steady-state operations. For such a task, it is critical to ensure that air (oxygen) from the regenerator section of the FCC is prevented from mixing with hydrocarbons in the reactor section. During normal operations, this is achieved by catalyst slide valves that are located between the regenerator and reactor sections and hold back a level of catalyst, and the catalyst circulation that counteracts the slightly positive differential pressure between the regenerator and reactor. However, during shutdown, when catalyst circulation and level are lost, this pressure differential must be reversed to prevent air in the regenerator from flowing to the reactor and the downstream plant that can contain hydrocarbons. In an FCC unit, this is normally achieved by maintaining a sufficient steam flow into the reactor to keep the pressure inside the reactor as the highest pressure point in the overall system. This critical steam barrier separates the air and hydrocarbon systems on either side of the reactor by flowing outward from the reactor, preventing the air and hydrocarbons from forming an explosive mixture.

During the FCC shutdown at the Husky Refinery, this, and other key safeguards were not implemented, and the operators were simply relying on the (nominally) closed catalyst slide valves to prevent air flow from the regenerator into the rest of the system. Slide valves were not intended to provide positive isolation and because they tend to wear in service, this further reduces their ability to prevent air from the regenerator from mixing with hydrocarbons in the reactor and downstream plant. This mixture found a source of ignition leading to the initial explosion.

Previous process hazard analyses (PHA) and layer of protection analysis (LOPA) studies did not consider transient operations, such as shutdown, that could lead to process upsets resulting in unintended flow of air from the regenerator into the reactor.

Per the CSB report, the following safeguards were not utilized:

- Reactor steam barrier: Husky Superior Refinery did not establish nor maintain a reactor steam barrier between air and hydrocarbons during the shutdown.

- Main column gas purge: Husky Superior Refinery did not properly purge the main column overhead receiver during the shutdown to remove oxygen from the system.
- Catalyst slide valves: Husky Superior Refinery relied entirely on catalyst slide valves to keep the air and hydrocarbon systems separated during the shutdown. Although refinery operators properly closed the slide valve, the valve had eroded and provided an open path that allowed a large flow of air to enter the reactor and migrate into the hydrocarbon-filled equipment, forming an explosive mixture.

Key Learning(s)

- The isolation-decontamination execution plans must fully consider published process technology knowledge to safely shut down the unit.
- Transient Operations, including shutdown, must be fully considered as part of the process hazard analysis (PHA) and Layer Of Protection Analysis (LOPA) studies.
- It is important to learn from the experience of other incidents; in this case, Husky Superior did not learn from a previous CBS report of a similar incident that occurred, involving another FCC unit transient operations explosion in California (CSB 2015).

6.3 COMPLETING AND DOCUMENTING EXECUTION OF THE PLAN

A formal document must show that the plan has been fully executed, according to the standards specified by the plan. Signatures on the plan document must clearly demonstrate that there have been no variations to the plan, or where changes have been made that they have been properly risk-assessed, authorized, and documented. Any changes to the plan should be carefully considered and involve those personnel who developed the original decontamination plan. Examples of situations where changes may be required include:

- Discovering that connection points such as drain valves and vent valves are plugged.
- Isolation valves are leaking by (passing) or are seized.
- Utilities required for purging are not readily available.

- Mechanical adapters for connecting to the chemical process may be incompatible and substitutes are needed.
- Blinds/blind flanges cannot be inserted in the planned location.

For turnarounds, the sequencing of activities is normally loaded into a master schedule that requires updating if any changes are made to the decontamination plan.

Oversight of the planned maintenance projects, the decontamination plans, and turnaround master schedule must all be considered and monitored by an operation planner or scheduler. Completing and documenting the status of each part of each decontamination plan for the facility is critical to the overall success of concluding the planned maintenance of the equipment and its return to service.

6.4 COMPLEX DECONTAMINATION PLANS

Some chemical facilities by their design are very complex with multiple pieces of equipment, extensive piping systems, and inter-connected unit operations. They may have recycle lines, common vent headers, waste treatment facilities, and common utilities. Decontamination plans for such facilities can take on an entirely higher level of complexity. Planning the decontamination of a chemical transfer line from the supply unit to the user unit, for example, would involve coordination between the unit decontamination planners, including discussions on how to purge the piping, when to purge, where the purge material is to be directed, when the piping is deemed clean, and the details on isolation methods and location. Another example is where toxic material is involved and it is difficult to drain all of the liquid material prior to purging, as described in Case Study 6-2.

Decontamination planning will require input from each of the affected unit operations and should be considered well before the planned shutdown. The plans requiring coordination between units will need more input and review from subject matter experts and management approval than for a single unit, as the experts and managers from all units must be made aware of the decontamination plans developed for other units.

Decontaminating equipment and piping to a common collection system, such as a central wastewater treatment plant, must also be carefully planned and communicated between units. The mixing of incompatible chemicals, for example, may create a reaction resulting in a safety or environmental event, so the facility must be aware of the timing and specific contents of the decontamination purge streams.

Case Study 6-2 — Decontamination of a Toxic Pipeline

Description

A pipeline containing a moderately volatile, toxic substance was being decontaminated by draining and purging to scrubbers. The final stage of the plan involved blowing compressed air through the pipeline to an open end and then testing the concentration of the substance coming out with the air until it fell below 20 ppm. When the decontamination target was achieved, workers stopped the air blowing and closed the open end of the pipeline. The next day, workers breaking into the line could smell the presence of the toxic substance and found that some of the material was still present in the line at dangerous levels.

Comments

Air blowing had been conducted at a relatively high rate, such that there was insufficient time for the substance to equilibrate in the air or to evaporate. Although the air leaving the system was below the required level, dangerous quantities of material were still present in low points of the pipeline.

Lessons Learned

- The decontamination plan must consider the physical properties of the material to ensure that it is removed to the required specification. In this case, using a lower purge rate of air as part of the decontamination plan would have shown that there were still higher levels of the material present.
- It is good practice to purge for a specific time period, stop the purge, and isolate the equipment for a period of time, then renew the purge and recheck the concentration. This sequence may have to be repeated several times.
- Suitable PPE for a first line break must be specified and used, even when decontamination is considered to have been completed.
- Retesting of isolated plant may be required at suitable intervals.

6.5 EQUIPMENT READINESS FOR ENERGY ISOLATION

As discussed in Chapter 5, decontamination and energy isolation can be an iterative process that is carried out in various phases. The integrity of the energy isolation may be relatively low during the early stages of the plan, where partial decontamination can be done, enabling a higher integrity of isolation such as insertion of a blind in a line. Once decontamination has been completed to the required standard specified by the plan, further energy isolation can take place, such as line removal and blanking.

6.6 SUMMARY

This chapter has highlighted the critical importance of having a documented decontamination plan that will ensure that decontamination is carried out to the required standard, can be tracked and checked by all interested parties (such as the example shown in Table 6-1), and any changes to the plan are documented/risk assessed before they are carried out.

Table 6-1 Decontamination Checklist

Guidance: *Review and check off the status of each checklist activity.*

Checklist Activity	Complete?				Checked by:		Comment
	YES	NO	N/A	Name	Date	Name	
Hazardous chemicals identified and communicated.							
Hazards associated with the substances used for purging or flushing are understood—asphyxiation, caustic/acid burns, high heat burns (steam), cold temperatures from rapid venting, pressure, chemical exposures, etc.							
Risk Assessment completed for the decontamination and isolation plan.							

Checklist Activity	Complete?				Checked by:		Complete?
	YES	NO	N/A	Name	Date	Name	
Environmental consequences evaluated such as flaring, venting, spillage, ground and water contamination, etc.							
Electrical isolation points identified and approved.							
Decontamination plan includes marked-up P&IDs, equipment drawings, isolation points, purge and drain connection points, and bleed/vent points.							
Required PPE documented and available.							
Emergency response plan in place.							
Decontamination plan documented and approved by SME's.							
Decontamination plan reviewed with affected operation and maintenance personnel.							
Decontamination plan reviewed with other affected personnel: emergency response; safety; contractors; etc.							
Decontamination plan reviewed with adjacent process units, common collection system/ wastewater plant, utility supplier, others.							

Checklist Activity	Complete?				Checked by:		Comment
	YES	NO	N/A	Name	Date	Name	
Simultaneous Operations (SIMOPS) activities reviewed, coordinated, and communicated to all affected personnel.							
Timing and conditions for implementing the decontamination plan discussed and agreed,							
If training was required, it was completed.							
Shift to Shift communication process and tool in place.							
Work Permits approved and in place, cross-referencing appropriate procedures and risk assessments.							
Walk the Line review completed.							
All isolation points are identified and match the P&IDs and isolation certificate							
All changes to the decontamination plan, have been reviewed applying MOC and approved at appropriate level.							

7 ISOLATION AND DECONTAMINATION PLAN EXAMPLES

This chapter includes examples of isolation and decontamination plans, with varying degrees of complexity, for both short and longer term, as well as plant and equipment mothballing/demolition.

7.1 ISOLATING/DECONTAMINATING LINES WITH TOXIC MATERIALS USING "DOUBLE BLOCK AND BLEED"

An example approach for isolating and decontaminating a piping system containing a toxic liquid chemical is described in this chapter and in Table 7-1 and illustrated in Figure 7-1. A similar plan could also be used for flammables, or material that is damaging to the environment. The basis for this example begins with the piping and valves that have been properly designed for isolation of hazardous chemicals using a double block, bleed and drain arrangement. In the diagram, the process area being prepared for maintenance is titled "Working Area" and the system remaining under pressure is titled "Pressurized system with liquid toxic chemical". The location of the maintenance is downstream (to the right) of Valve V2 shown in Figure 7-1.

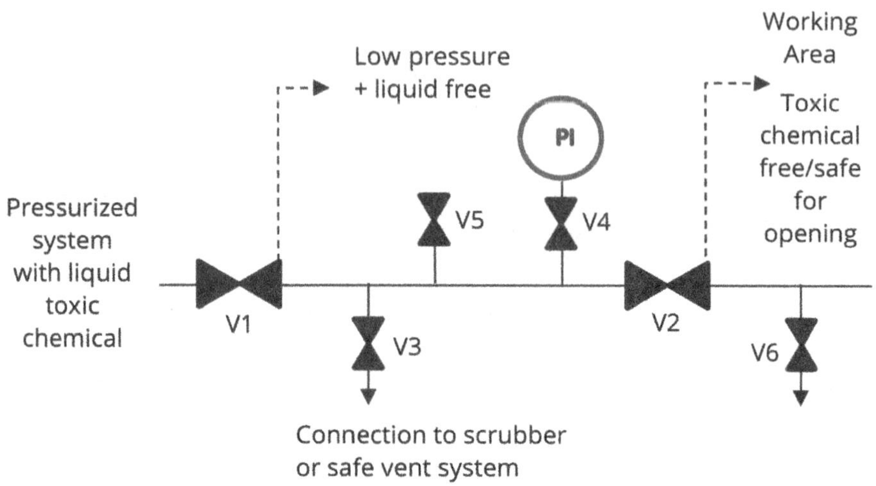

- Working Area is downstream of the two closed valves (V1 and V2)
- Pressure gauge (PI) between V1 and V2 to indicate liquid free/low pressure (continuously indicated)
- Area between V1 and V2 provides the ability to depressurize the system with a temporary or fixed connection to a safe location.
- V3 is normally closed after depressurization is complete. V4 is normally open for pressure indication.
- V5 is a purge connection for purging area between V1 and V2 and can also be used for purging piping downstream of V2 (once V1 has been proven an effective isolation).
- V6 is a purge connection on the equipment to be worked on.

Figure 7-1 Isolation Design for a Piping System Containing a Toxic Liquid Chemical

The general steps to follow are detailed in Table 7-1.

Table 7-1 Isolation and Decontamination Steps for Hazardous Material

#	Step	Notes and Alerts
1	Test and calibrate the pressure gauge (PI).	Pressure gauges are often out of calibration, stuck, or may be plugged.
2	Select appropriate Personal Protective Equipment (PPE)	Assume residual material will be present for first line break.
3	Close valve V2 and allow process area downstream of V2 to depressurize. Monitor pressure reduction in downstream equipment to determine if V2 provides positive isolation.	Ensure any pressure instrument is properly calibrated.
4	Close and lock V1 and open V2 so the piping between V1 and V2 can de-inventory to the user vessel	V1 should be locked, or car sealed and tagged per risk assessment.
5	Close and lock V2	V2 should be locked, or car sealed and tagged, per risk assessment.
6	While wearing appropriate PPE, carefully remove the plug or blind flange to V3 (normally closed).	The space between the valve and the plug/blind flange may be pressurized if V3 is leaking by.
7	Connect temporary drain piping or tubing between V3 and a safe disposal system (such as a scrubber)	Check that the pressure rating and materials of construction of the temporary piping or hose are suitable for the duty.
8	Slowly open V3 to depressurize the piping between V1 and V2	Take care not to overload the vent system or scrubber.
9	Monitor the PI pressure	Pressure should fall steadily to zero.

#	Step	Notes and Alerts
10	If pressure does not drop, valve V3 or line may be plugged.	Reevaluate plan and check or provide additional risk assessment, e.g., for use of rodding procedure to clear blockage.
11	After the pressure gauge shows zero, close V3 and check the piping between V1 and V2 after a short period of time to see if the PI shows pressure. If pressure is observed, then likely V1 is leaking by (passing). If no pressure is observed, then V1 is providing a tight shutoff.	For liquefied gases, if they are vented too quickly, they may auto-refrigerate and form pools of liquid which subsequently evaporate and pressurize the system. This can be confused, as the symptoms are similar to V1 leaking by.
12	If V1 is holding, connect a purge line to V5, open V5 and V3 to purge the area between V1 and V2 to the safe disposal system.	Check that the pressure rating and materials of construction of the temporary piping, tubing, or hose are suitable for the duty.
13	After purging, close V3, open V6 and V2 to purge the piping to be worked on via V6.	Alternatively, could purge back from the equipment to be worked on via V5.
14	The purge stream on the work area must be frequently checked (for pressure, contamination level) until considered safe for line break.	Pressure/recontamination could build up due to valve leaking by, backflow from vent line, vaporization of liquid legs.
15	Once the line is considered safe, turn off the purge, close V5, V2 and V6.	Valves should be locked, or car sealed and tagged per risk assessment.
16	Continue to monitor the pressure between V1 and V2	To check for valve leaking by.
17	After the line break is completed, maintenance of the piping can be initiated.	

If V1 or V2 is found to be leaking by, then an alternative isolation and decontamination plan must be created using MOC and a revised risk assessment.

Note that if a double block and bleed arrangement was not available, Valve 1 should not be considered adequate for isolating a highly hazardous toxic or flammable material. The leaking by of small amounts of chemicals could pose unacceptable exposure or ignition conditions for personnel performing the maintenance task. Conducting a thorough risk assessment is always recommended if a single isolation device against a hazardous substance is being considered. Under such circumstances, it may be safer to consider a wider scope of isolation plan that achieves the required level of isolation.

7.2 ISOLATION/DECONTAMINATION FOR VESSEL ENTRY

Where the work to be conducted involves personnel entering a confined space such as a reactor or distillation column, a higher standard of isolation and decontamination should be considered. Initial isolation and decontamination may follow a similar path to that detailed in 7.1, but most facilities will require physical isolation from sources of energy as well as a higher level of authorization within the organization. Energy Isolation is discussed further in Chapter 8 and 9.

An example of a decontamination procedure for entering a vessel that contained toxic and flammable material, once liquid draining/physical isolation has been achieved, is provided in Figure 7-2 and Table 7-2. This example is challenging because there is no route to purge air through the vessel inlet to its outlet. Asphyxiation can result, as discussed in Case Study 7-1 (CSB 1999).

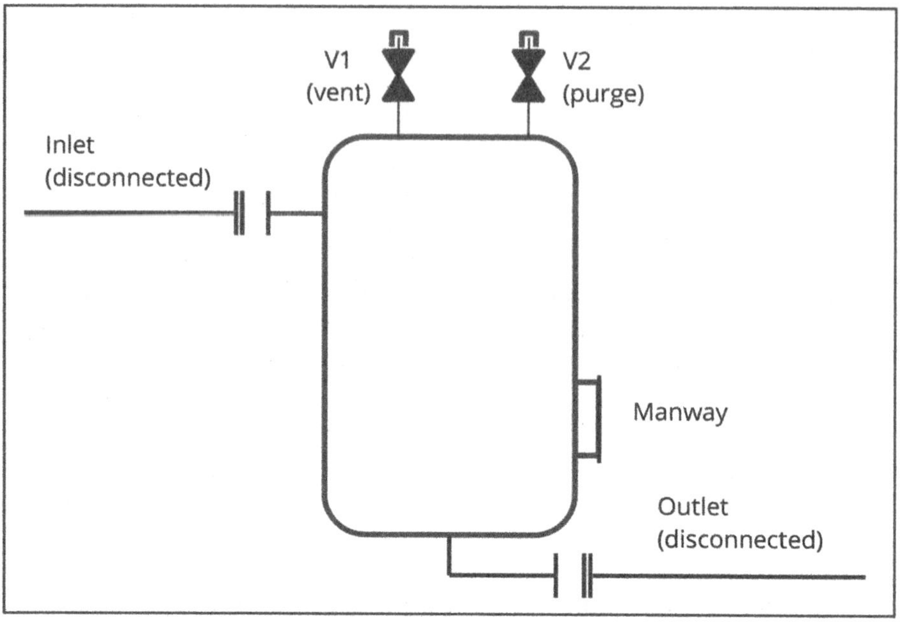

Figure 7-2 Vessel Isolation Schematic for Confined Space Entry

Table 7-2 Decontamination for Vessel Entry

#	Step	Notes and Alerts
1	Verify vessel has been drained and physically disconnected per current procedure XYZ.	Check latest procedure was used, verify signatures and discuss with team who conducted isolation.
2	Remove plug/blind from V1 (vent) and V2 (purge)	PPE for first line break in case of material trapped between valve and plug/blind flange
3	Connect V1 to vent line via dedicated vent hose to vent scrubber	Vent and Purge hoses are color-coded, and connections are specific to each hose type to avoid mixing them up.

#	Step	Notes and Alerts
4	If purging with nitrogen, check nitrogen system pressure does not exceed 10 barg using calibrated gauge and adjust regulator if required	Design pressure of vessel is 15 barg and must not be exceeded as pressure relief is on downstream piping and is therefore isolated.
5	Connect V2 to nitrogen purge header via dedicated purge hose	Nitrogen is an asphyxiant; do not allow nitrogen venting when personnel are present. Signs explaining the dangers of nitrogen to be displayed and barriers erected.
6	Open V2 and slowly pressurize vessel to 5 barg	Monitor for leaks from blanked feed and outlet lines.
7	Close V2 and slowly open V1 to depressurize vessel to atmospheric pressure	Monitor vent pressure; do not overload vent system.
8	Repeat (6) and (7) three times	Remove residual flammable material before air is introduced.
9	When vessel is at atmospheric pressure, close V1 and disconnect nitrogen hose	
10	Connect plant air hose to V1 and repeat (6) and (7) twice	Repeated pressurization with air will remove sufficient nitrogen to be safe to open the manway. Verify that plant air is not cross connected with process air which may be nitrogen.
11	Sample from the vent line and test for toxic/flammables/air	Check if within safe limits.
12	If analysis is within specified limits, manway can be removed. Remove inlet and outlet pipe spool and fit blank on feed and outlet piping leaving open pipe on vessel nozzles.	PPE for fitters—in case low-level contamination is still present. Vessel is physically disconnected from all feed and outlet lines.

#	Step	Notes and Alerts
13	Sample and test gas in vessel contents from lower manway. If flammable levels are above 10% LEL or toxic material above exposure limits as defined on Safety Data Sheet, replace manway and repeat steps (5) to (13).	DO NOT put your head near manway in case toxic material/nitrogen is present.
14	Install >500 m^3/h (300 cfm) air mover in bottom manway and run for 8 hours	
15	Sample and gas test vessel at top and middle (sample tubes inserted through the vent/purge valves) and bottom.	Check substances are within exposure limits as defined on Safety Data Sheet and oxygen is above 20%.
16	If analysis is within specification for entry, carry out internal inspection using floodlight from outside the manway.	Check for visible internal deposits that might release hazardous material, if disturbed.
17	Complete confined space vessel entry permit with repeat gas testing every hour or continuous monitoring if possible, or as required by regulations/company standards.	Standard procedure to have standby person with radio and emergency escape equipment in visible contact with person inside vessel. Standby person not to enter vessel but to call for support if required.

Case Study 7-1 —Nitrogen Asphyxiation

Description

During a turnaround, two workers, who were inspecting process equipment for residues using ultraviolet (UV) light, were found collapsed inside a makeshift tent. One died and the other eventually recovered.

Comments

New catalyst had just been installed and nitrogen was being injected into the process equipment to protect the catalyst from damage due to exposure to moisture. The nitrogen was venting to atmosphere via a 48" pipe and the workers were inspecting the flange surfaces of the 48" pipe with the UV light. Daylight made the task difficult, so they placed a black sheet over the pipe to create some shade, inadvertently creating a confined space where the nitrogen accumulated, asphyxiating them.

Key Learning

The hazards of working with nitrogen are often not properly understood. Nitrogen is colorless, odorless, and tasteless. Workers who use nitrogen require formal training on its hazards.

Activities involving nitrogen where workers are present must be very carefully controlled via a thorough risk assessment. Precautions include restricting access, clear signage, and workers wearing personal gas monitors.

See Section 3.3.1.1 for more details about safe use of nitrogen.

7.3 ISOLATION FOR LONGER TERM/MOTHBALLING

Where equipment is expected to be out of service for an extended period of time, additional considerations are required. Best practice is for full decontamination and physical isolation (air-gapping) of the equipment, i.e., similar to that required for vessel entry described in 7.2. While use of a blind between two flanges might be considered appropriate for a relatively short period of time, since it is not possible to inspect the condition of the blind, this should not be used for longer term isolation. Long-term isolation may result in dead-legs on the service side up to the point of isolation, which may result in corrosion. This asset integrity issue must be considered by subject matter experts.

Preservation of the equipment including associated instrumentation and relief systems must also be considered, as discussed further in Section 12.7, Delayed Return to Service.

7.4 ISOLATING/DECONTAMINATING BUTYL LITHIUM STORAGE SYSTEMS

The unique properties of butyl lithium warrant a number of additional steps for isolation and decontamination compared to those required for a storage system containing a more typical toxic or flammable material.

7.4.1 Understanding the Hazards

Lithium alkyls (e.g., butyl lithium) are used in fine chemical synthesis or as catalysts in anionic polymerization (for example, the production of SSBR rubber used in car tires). They are produced by the direct combination of lithium metal with the corresponding alkyl halide.

Butyl lithium is temperature sensitive, decomposing at temperatures above ~30 °C to produce solid lithium hydride (LiH) and butene gas. The butene is largely soluble in the carrier solvents typically used with butyl lithium, so little or no pressure increase is noted. However, the *flashpoint of the carrier solvent is considerably reduced*. If allowed to settle, LiH deposits will become hard and resemble marble in appearance. Solid deposition can occur in overheating process equipment, or in stagnant sections of pipework, but is a particular problem in storage tanks due to the longer residence time.

Pure butyl lithium is not strictly pyrophoric but is *air reactive* (forms lithium butoxide and heat, which accelerates the decomposition process). The pyrophoric nature usually observed is driven by the presence of finely divided LiH, which *rapidly generates hydrogen and heat* upon contact with atmospheric moisture.

Hazards for maintenance are therefore flammable hydrocarbons and a pyrophoric mixture of lithium alkyl and LiH which easily liberates hydrogen gas. Additional hazards to personnel include thermal burns from the solvent and chemical (caustic) burns from the lithium hydroxide (LiOH) produced during the combustion. Inhalation of combustion products carries some risk from the LiOH fumes. The standard decontamination procedure uses "reactive cleaning" with water or an alcohol to remove solid deposits of LiH. "Reactive cleaning" uses the inherent properties of the materials to react chemically with the cleaning material. This results in the release of hydrogen gas, which has to be vented safely. It may be necessary to introduce a purge of inert gas (e.g., nitrogen) to maintain hydrogen levels safely below the lower explosive limit. The significant amount of heat generated during reactive cleaning must be managed to avoid over pressurization or other equipment damage.

An example method is provided in Section 7.4.2.

7.4.2 Isolation and Decontamination- Butyl Lithium Storage Systems

1. Isolation of the equipment can be complex. Due to the pyrophoric nature of the material, blinding or creating air breaks may not be possible. The injection of clean, dry mineral oil into surrounding lines via double-valved purge lines (situated at the top of process lines, to avoid blockage with solids) can be used to somewhat mitigate the pyrophoric nature of material that could present a hazard during an initial line break.

2. Drain all possible liquid from the system and purge with nitrogen or other inert gas (argon is the alternative, as butyl lithium reacts with CO_2). Do not blow the system dry during this step, as the evaporation of the carrier solvent will increase the decomposition rate of the butyl lithium and formation of LiH deposits.

3. Flush the system with clean, dry hydrocarbon solvent (typically the same as the carrier solvent for the butyl lithium solution) to dilute and remove any remaining butyl lithium. This process is repeated until the flush solvent is no longer air reactive (typically <5% w/w butyl lithium and preferably < 1%). A simple test for residual butyl lithium content is to hydrolyze a known volume of the wash solvent and measure the pH of the resulting aqueous LiOH solution, assuming ideal behavior of the LiOH solution.

4. Provided it is possible to vent the equipment, the next step is to flush with damp hydrocarbon solvent (i.e., only containing dissolved water). For most solvents, this typically is about 100-200 ppm moisture. The effluent solvent is analyzed for moisture and this measurement is used to give a qualitative indication of the presence of LiH within the system.

5. If the system cannot be safely vented or is small (typically process pipework below 3" diameter), it may be necessary to omit the use of water and move directly to using oil passivation. If the levels of LiH are believed to be significant, then every effort should be made to use reactive cleaning with water first. The use of a zoned endoscope or camera should be considered to make a preliminary internal inspection prior to the addition of water.

6. As water contacts LiH deposits, a layer of lithium hydroxide (LiOH) is formed. While LiOH has reasonable solubility in water, the layer formed can have a protective effect on the deposit of LiH beneath, such that the LiH deposit may not be fully reacted. The presence of a stable moisture level in the effluent solvent cannot therefore be taken as proof that all the LiH has been reacted. Note that adding water should be suspended if the vessel temperature increases by more than 20 °C (note – it can be difficult to obtain representative temperature readings during this stage).
7. LiH deposits are usually formed at low points, typically in the base of a storage vessel. The final stage of decontamination is carried out by slowly introducing water to the vessel while circulating a suitable quantity of solvent (which acts as a provider of turbulence and as a heat sink). Adequate venting must be provided and monitored during this operation; typically, the reaction between LiH and water can be considered as effectively instantaneous. Typically, once the bottom head or lower surface of the vessel is covered with water, the addition rate can be increased until the vessel is full. It should then be held under a nitrogen blanket for 48 hours to allow any residual reaction to take place.
8. The nitrogen blanket is maintained while the water is drained. Check the pH of the effluent water as this gives an indication of the quantity of lithium-containing species that were present. The vessel vent should be monitored for flammability prior to cautiously introducing air into the vessel and preparing for entry. Any remaining solid deposits should be sprayed with clean mineral oil prior to attempting to disturb them.

NOTE: On small systems, or where venting is difficult (Step 4), the system should be flushed with clean dry mineral oil. The oil forms a boundary coating over the LiH deposits, which reduces the rate of hydrogen production when in contact with moisture and reduces the probability of ignition. If the line is largely blocked, the oil may be pumped in under pressure and allowed to slowly soak through the deposits. After this, line breaks should be carried out with the expectation that local fires will occur and so appropriate precautions should be in place (extinguisher and standby provision, metallized Nomex or similar PPE, etc.).

7.4.3 Return to Service- Butyl Lithium Storage Systems

Once maintenance work is completed, the system must be thoroughly purged with nitrogen prior to drying (typically O_2 <5%). The standard approach is then to wash with clean, dry solvent followed by a very dilute solution of butyl lithium in the carrier solvent (typically 1-5% w/w). This will remove any residual traces of water or oxygen without producing significant levels of solid formation.

7.4.4 Key Points- Butyl Lithium Storage Systems

- Dealing with the *typical hazards* from isolating and deisolating.
- The *progressive removal of hazardous materials* – first the alkyl lithium solution, then the reactive deposits – but in the expectation that they cannot be removed entirely.
- *Passivate remaining deposits* with a mineral oil coating.
- *Fully drying equipment* as part of the return to duty is important, both for reasons of safety and quality.

A typical checklist is provided in Table 7-3 and a sample return to service checklist is presented in Table 7-4.

7.4.5 Example High-level Checklist for Butyl Lithium System[2]

Table 7-3 Isolation and Decontamination Checklist

Step	Description	Hazards and Mitigations	Check
1	Isolate the equipment by blinding or creating air breaks (if possible), by using mineral oil to mitigate the pyrophoric nature of any exposed material.	Releases of flammable/ pyrophoric liquids. Material is drained into clean, dry mineral oil in a steel container. Drained materials are then quenched remotely.	
2	Drain the system and blow with inert gas to minimize the presence of pyrophoric liquids.	Minimize the presence of pyrophoric inventory and maintain an oxygen and water-free environment.	
3	Flush repeatedly with a compatible solvent until washings are non-reactive.	Confirm removal of all soluble reactive material.	
4	Flush with damp solvent to quench reactive material, measuring the water content of the effluent solvent to monitor the process.	Controlled reactive cleaning of residual reactive material. Monitor progress by monitoring temperature (e.g., IR) and vent gas flow.	

2 This example checklist is provided for illustration only.

Step	Description	Hazards and Mitigations	Check
5	Quench reactive residues by introducing water at a controlled rate, such that the gas evolution can be safely vented.	Controlled reactive cleaning of residual reactive material. Monitor progress by monitoring temperature (e.g., IR) and vent gas flow.	
6	Flood the system with water and monitor for gas evolution over at least 24 hours.	Allows time for any hardened deposits to react out. LiOH deposits that form can protect underlying LiH deposits from hydrolysis for some time.	
7	Purge the system with nitrogen while draining the wash water.	Prevents sudden ingress of air until cleaning status is confirmed.	
8	Check pH of effluent water.	Gives guidance on the quantity of lithium-containing species that remain in the system.	
9	Monitor vent gases while progressively introducing air to the system.	Check LEL of exit gases before reaching 8% volume of O_2. Check for flammability.	
10	Coat any remaining solid deposits with mineral oil before attempting mechanical cleaning.	LiOH deposits that form can protect underlying LiH deposits from hydrolysis for some time. Mineral oil helps passivate remaining LiH deposits.	

Table 7-4 Return to Service Checklist

Step	Description	Hazards and Mitigations	Check
1	Inert the system.	Removes oxygen and helps dry the system.	
2	Fully dry using solvent flushing or repeated purging.	Removes any residual water. Check effluent solvent for water content using Karl Fischer analysis.	
3	Introduce a very dilute solution of the lithium alkyl to scavenge any remaining water or oxygen.	Confirms dryness and anaerobic conditions prior to introducing product to the system.	
4	Drain while maintaining an inert gas purge.	Removes material/contaminants that would result in out-of-spec product.	
5	De-isolate.	Returns equipment to service.	

7.5 EXAMPLE – ISOLATING/DECONTAMINATING A TOLUENE DIISOCYANATE REACTOR TRAIN

7.5.1 Description: Toluene Diisocyanate Process

This example provides a high level review of an isolation and decontamination plan on a hazardous and relatively complex process and would form a part of a wider procedure for the preparation for maintenance and return to service,[3] which would include a detailed Energy Isolation Plan (EIP), leak testing, recommissioning, and a startup plan.

The toluene diisocyanate process is designed with three identical reactor trains, each consisting of reactant feed, continuously stirred tank reactor (CSTR), phase separator, and kettle feed preheater for the downstream HCl column, as shown in Figure 7-3.

The main reaction chemistry is as follows:

Toluenediamine (TDA) + phosgene → toluene diisocyanate (TDI) + HCl + residues

$$CH_3C_6H_3(NH_2)_2 + 2COCl_2 \rightarrow CH_3C_6H_3(NCO)_2 + 4HCl + \text{residues}$$

A 6-week rotating maintenance schedule is required on each of the three reactor trains to remove solids formation and fouling in the equipment by circulating a soapy water solution through the process.

[3] This example is for illustration only.

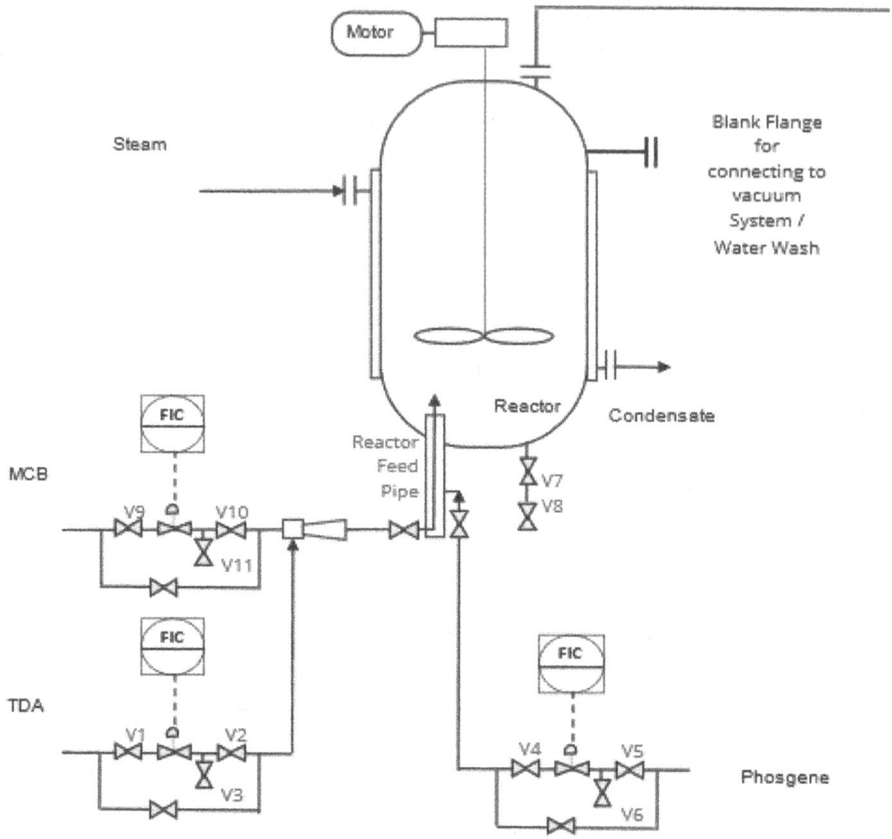

Figure 7-3 Diagram of TDI CSTR System (1 of 2)

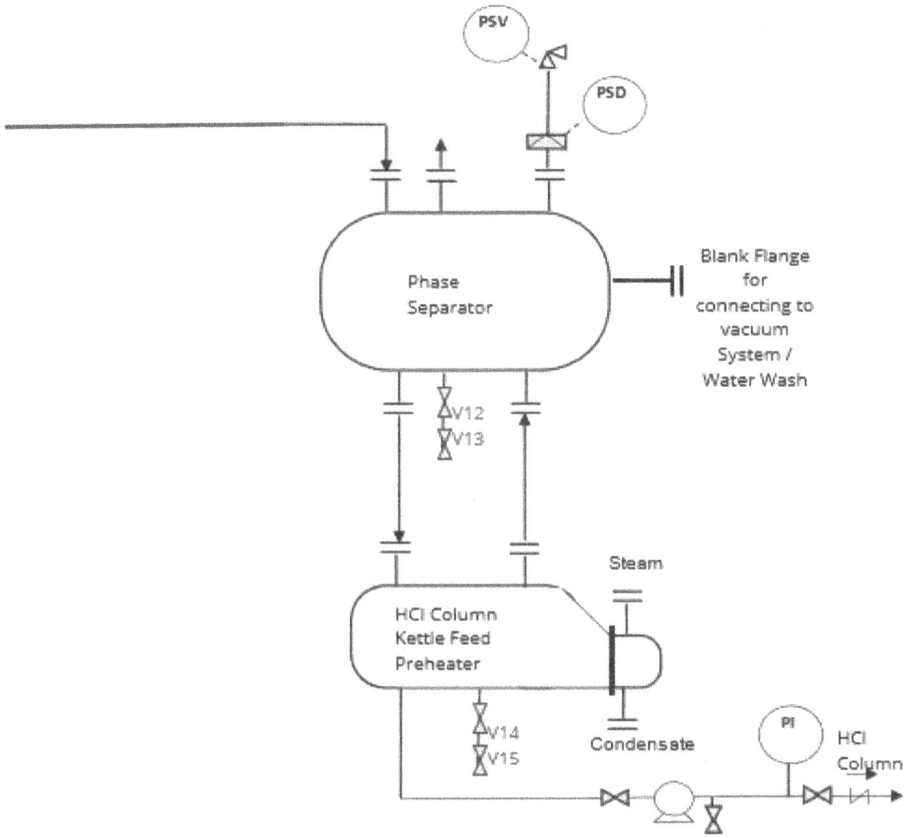

Figure 7-3 Diagram of TDI CSTR System (2 of 2)

7.5.2 Hazards of the Process:

The chemicals involved in the process include toluene diamine (TDA), a suspected carcinogen and mutagen; monochlorobenzene (MCB) a flammable liquid; phosgene ($COCl_2$) a toxic inhalation hazard; anhydrous hydrogen chloride (HCl) a toxic inhalation hazard; toluene diisocyanate (TDI) a toxic liquid, and various urea and polymerized isocyanate solids.

7.5.2.1 Chemical Hazards

The chemical hazards can be summarized as follows:

- Phosgene and HCl are highly toxic gas with IDLH (Immediate damage to life and health) of 2 and 50 ppm, respectively, and TLV (Threshold Limit Values) of 0.1 and 2 ppm, respectively.
- TDI is a toxic liquid and can be fatal if inhaled with TLV of 0.001 ppm, and STEL of 0.005 ppm.
- TDA is solid at room temperature and toxic if swallowed or in contact with skin.
- Phosgene and HCl are water reactive, immediately forming muriatic acid (hydrochloric acid solution). The resulting reactive products are highly corrosive, especially to carbon steel equipment.
- TDI is water reactive, forming solid urea with CO_2 liberation. Excessive CO_2 liberation could lead to equipment overpressure.
- MCB is a flammable liquid with a flash point of 75 °F (24 °C), LEL = 1.3 vol% and UEL = 7.1 vol%
- The solids produced in this process can experience an auto-oxidation and subsequent fire if exposed to air for a period of time. The autooxidation potential is increased with increasing temperature.

7.5.2.2 Equipment Hazards

The Equipment Hazards can be summarized as follows:

- Phosgene can impregnate metal equipment and is subsequently liberated under high heat situations, such as welding.
- Solid reaction products tend to block low-point bleed/drain valves that require rodding out before equipment can be properly drained.
- The design pressure of the reactor, phase separator, and kettle is 300 psig.

7.5.3 Decontamination Plan

This decontamination plan takes about two days to complete working between 2 × 12-hour shifts. Operational continuity plans include:

- Written instructions for the oncoming shift.
- Responsible operators shift to shift notes with details on work completed during the shift.
- Shift handover/relief with decontamination plan in hand in the field.
- Shift relief meetings discussing the progress to the plan.

Given the hazards and complexity of this plan, a checklist standard operating procedure (SOP) is used with independent verification of completion of each step as written. A sample Isolation/Decontamination plan is presented in Table 7-5. Deviation from the plan would require management of change (MOC).

Table 7-5 Outline Standard Operating Procedure for Isolation and Decontamination

Step	Instruction	Comment
1.	Stop TDA feed pump, isolate control valve using upstream and downstream block valves (V1 & V2), leaving control valve open to allow draining of liquid between the block valves.	Bypass valve to remain closed. Note: The presence of the bypass valve means isolation is only against a single valve at this stage.
2.	Drain TDA control valve leg via drain valve V3 to toxic drain tank	
3.	Stop phosgene feed, isolate control valve using upstream and downstream block valves (V4 & V5), leaving control valve open.	Bypass valve to remain closed. Note: The presence of the bypass valve means isolation is only against a single valve at this stage.
4.	Connect vacuum system to phosgene control valve leg via valve V6 and vent to caustic scrubber	Verify phosgene has vented down via calibrated pressure gauge.

Step	Instruction	Comment
5.	Close control valves in steam supply to reactor and kettle and isolate manual valves in steam and condensate lines.	
6.	Note: Continue MCB feed to reactor for 10 hours at a rate of x kg/h	MCB flush will dissolve most of the phosgene, HCl, and TDI for processing in solvent recovery
7.	Rod out bleed valve V3 from TDA reactor feed line using Reactor Rod Out Procedure SOP XX and rod-out tool	SOP XX includes: (1) Specified PPE for first line-break; (2) careful removal of bleed valve plug, while capturing fugitive emissions with vacuum hose; (3) screw in rod-out tool to bleed valve; (4) open bleed valve and rod-out; (5) retract tool and close valve; (6) unscrew rod out tool
8.	Stop MCB feed pump and isolate control valve by closing upstream and downstream block valves V9 and V10. Leave control valve open to allow draining of liquid between the block valves.	
9.	Drain MCB control leg via drain valve V11 to Impure Solvent Tank	
10.	Install temporary copper tubing from phosgene control valve bleed valve (V6), reactor drains (V8), phase separator (V13), and kettle (V15) to impure MCB tank.	
11.	Open bleed valves and drain down to impure MCB tank. If copper tubing plugs, remove per first line break procedure and replace.	Time to drain may be variable but should be complete in ~6 hours. Used/plugged tubing to be disposed of via hazardous waste procedure

Step	Instruction	Comment
12.	Implement Lock-out/Tag-out system per Energy Isolation Plan (EIP). See note *.	EIP Steps, possibly in checklist format (not included in this example)
13.	Remove blanks from flanges at top of Reactor and Phase Separator and connect to vacuum/scrubber system via nozzle.	First Line Break PPE requirements
14.	Pull vacuum on Reactor and Phase Separator for 1 hour	Reduce toxic concentrations to near the IDLH values
15.	Remove PSV and bursting disc	To prevent damage by washing process
16.	Connect wash water to reactor and phase separator via top flange nozzle (where the vacuum is pulled) and wash to chemical sewer	Water wash for approximately 6 hours, then drain reactor
17.	Remove reactor feed pipe per maintenance plan	Physical isolation from all feed
18.	Connect nitrogen hoses to low point bleed valves on reactor, phase separator and kettle	
19.	Add soapy water and open low point bleed valves to allow nitrogen to sparge and mix soapy water and vent to vacuum system. Soapy water wash reactor and phase separator system per SOP for approximately 12 hours	Carefully monitor water levels to ensure water does not enter vacuum system.
20.	Monitor vapor space of reactor and phase separator for LEL and phosgene, using phosgene badge, draegar tubes and LEL monitors.	Ensure no toxic/flammable mixtures are being generated. Recommend every 2 hours
21.	Isolate soapy water supply and drain down vessels and piping	Decontamination and washing complete. Refer to SOP XYZ for drying and preparing for return to service.

*Note: In this example, if vessel entry for cleaning were required, further isolation of energy sources would be necessary and detailed on the energy isolation plan (EIP) including, but not limited to:

- Physical disconnection of inlet/outlet process piping to vessels
- Physical disconnection of steam and condensate lines
- Removal of fuses from pump and agitator motors and locking off electrical supplies
- Removal of radioactive sources

7.6 SUMMARY

This chapter provides some examples of isolation and decontamination plans and procedures, including a brief discussion on longer term isolations. As demonstrated in the case examples in this chapter, isolation and decontamination can be a complex task that requires multiple procedure steps and trained personnel. Failure to follow the steps per the written procedure, making a human error, or miscommunication between personnel may create an unsafe isolation condition.

8 OVERVIEW OF ENERGY ISOLATION

This chapter provides an overview of the various types and degrees of energy isolation and their relative security. It discusses the methods for isolating different energy sources as well as some examples of the equipment that is used as part of a Lock-Out Tag-Out (LOTO) procedure.

Chapter 3 described the various types of energy hazards that can be encountered when isolating and decontaminating process plant for maintenance including:

- Chemical
- Pressurized liquids and gases
- Thermal
- Electrical
- Kinetic
- Potential
- Radiation

Energy Isolation Plans (EIPs) including more specific isolation requirements are discussed further in Chapter 9. Energy isolation is also discussed in the online CCPS book on energy isolation (CCPS 2022c) and further guidance is available in the UK HSE guide on the safe isolation of plant and equipment (HSE UK 2006).

8.1 DEGREES OF ISOLATION

The degree, or level of isolation required must be assessed based on the acceptability of the level of risks that remain to the facility, equipment, personnel, or the environment. The expected period of isolation also requires consideration, since it may be acceptable to have a lower level of isolation for a very short period, when extra controls can be implemented and tightly controlled, but this may not be appropriate for work involving a longer duration. Isolation could range from closing a single valve or electrical switch to physically disconnecting the source of energy (positive isolation) and applying multiple locking devices. Ideally, at least one of the isolations should be as close as possible to the maintenance point or work zone.

A risk assessment is recommended to determine the most appropriate type and degree of isolation that could range from, in order of rigor:

- A local discussion based on historical process knowledge and experience, for low-risk isolations, such as water lines.

- A more structured team analysis and the development of a written isolation plan, including LOTO.
- A structured risk analysis that could involve using recognized Hazard Identification and Risk Analysis (HIRA) tools such as Hazard and Operability Analysis (HAZOP), what-if/checklist analysis, Bow-Tie Analysis, Failure Mode and Effects Analysis (FMEA), Failure Modes, Effects and Criticality Analysis (FMECA), and even completing a Layer Of Protection Analysis (LOPA) for consequences posing the highest risks.

Human factors should be considered as a key part of the Energy Isolation Plan as detailed in 8.2.

If there are doubts about the HIRA tools to utilize, request the Involvement of subject matter experts (SME) to determine if a HIRA is required and the appropriate tool to utilize. Various SMEs may be required if the process to be isolated poses multiple hazards such as chemical, electrical, mechanical, or radiation.

Some examples of the relative security of various isolation types are provided in Table 8-1, which would typically be used as part of a procedure for locking and tagging the isolation. A robust Lock-Out Tag-Out (LOTO) procedure is not only good engineering practice, but is a legal requirement in many jurisdictions, particularly in the USA, see Section 8.4.

Table 8-1 Relative Security Level of Isolation Types for Different Energy Sources

Energy Source/ Relative Security	Low	Medium	High
Process Isolation (Chemical, pressurized fluids, thermal)	Closed Valve	Double block and bleed	Positive isolation such as physical disconnection of piping
Electrical	Switch off	Switch off and locked	Fuses or breakers removed
Kinetic	Switch off equipment	Switch off and tie-off (e.g., cooling tower fan)	Physically disable such as disconnecting drive or removal of equipment
Potential	Tie off equipment (e.g., where it can fall)	Lock equipment in place	Remove equipment
Radiation	Signage and barrier for maintaining distance of personnel	Close and lock radiation shield (e.g., level instrument)	Move radioactive source to safe area using shielded device

Further details of isolation from energy sources are provided in Chapter 9.

8.2 HUMAN FACTORS IN ENERGY ISOLATION

Many of the human activities conducted at facilities that involve stored energy are prone to human failure. Energy isolation is a key area where a mistake could lead to unplanned release of energy resulting in injury or damage to the facility or the environment. A key purpose of many of the steps taken in energy isolation is to reduce potential for human error by:

- Conducting a risk assessment.
- Developing a documented plan and procedure.
- Communicating the plan to the team.
- Labelling and locking isolation points.
- Walking the Line.

Where the potential consequences are high, studies such as Safety Critical Task Analysis (SCTA) can identify activities where there is the potential for human error that could lead to an incident, allowing appropriate mitigations to be put in place. Action arising from a human factors SCTA may include:

- Improved labelling/ notices.
- Improved lighting.
- Consideration of fatigue/ workload.
- Added supervision.
- Development of memory aids such as a checklist.
- Dual sign-off of procedural steps.
- Additional independent barriers such as disconnecting valve actuators.

Further guidance on human factors is provided in the CCPS Handbook (CCPS 2022a) and additional information on human factors SCTA is available from the Energy Institute (EI 2020).

8.3 DESIGNING FOR ENERGY ISOLATION

The requirements for energy isolation should be considered at the design stage, particularly where frequent isolation is expected, such as where equipment has to be routinely removed for maintenance. Examples include a double block and bleed, a spectacle blind that can be rotated to open or close (see Figure 8-1, (Wermac online), fast acting line blinding system (see Figure 8-2, (F Corporation online), removeable pipe spools, easily accessible and lockable electrical isolators (see Figure 8-3 (PHAM News online). Such systems should be designed to avoid the potential for human error, including clear and unambiguous labelling and risk assessed/ documented procedures.

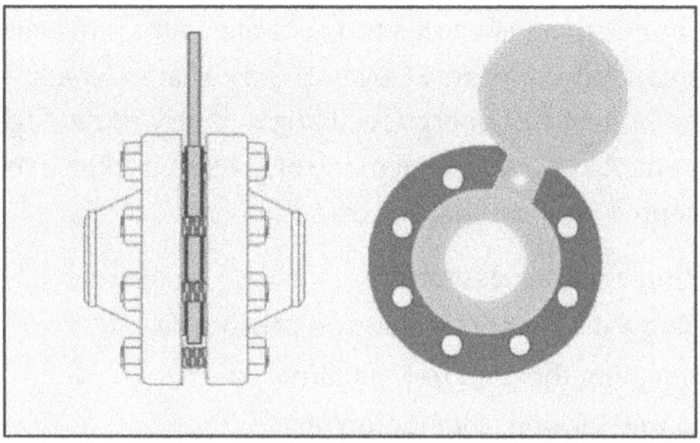

Figure 8-1 Spectacle Blind Shown in Open Position

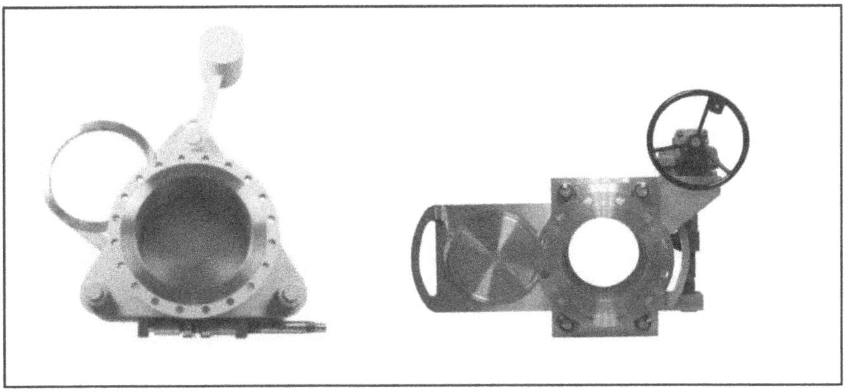

Figure 8-2 Fast Acting Blind Valves

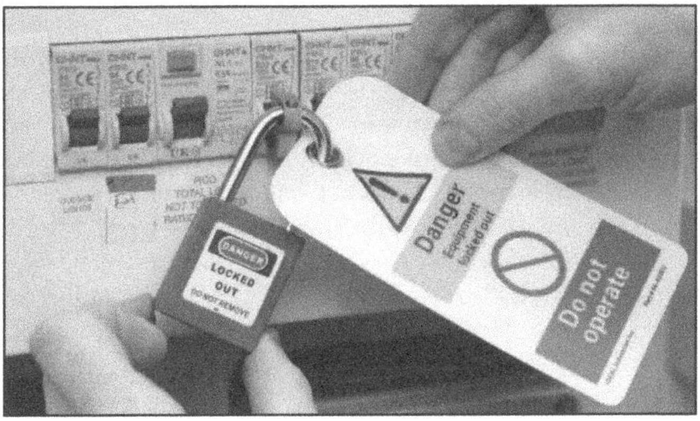

Figure 8-3 Electrical Isolator Lock

However, particularly with an older plant, it is not always possible or practical to utilize such energy isolation systems, leading to a requirement to develop appropriate plans using what is available on the facility. For example, where an unproved single isolation is available, it may be necessary to conduct a carefully controlled first line break to install a blind between flanges. This would require a carefully prepared and documented procedure that would include (but not be limited) to the following:

- Senior management approval including approved deviation from existing safe management practices.

- Subject Matter Expert to check that the design of any blinds/blind flanges, including thickness and materials of construction, is consistent with piping specification and expected substances and pressures.
- Wearing a full chemical suit and breathing apparatus (for worker and standby personnel).
- Use of a gas monitor.
- Barriers to keep non-essential personnel away from the area.
- Use of safe break-in technique (undoing bolts on opposite side of flange, staying out of the line of fire).

Under some circumstances, it may not be possible to complete safe isolation on part of the process and the work will require a pause until the entire facility is shut down and suitably decontaminated.

8.4 LOCKING AND LABELLING

Once the isolation has been carried out, a system of locking and labelling is required. Depending on the level of risk, that should be established via a risk assessment, this could range from applying a simple labelled tag to a piece of equipment to a formal system of Lock Out Tag Out (LOTO), which includes labelling a tag associated with a padlock, plastic, or wire loop. The label should have a reference number and a note to identify the reason and person or department responsible for the point of isolation. It must be recognized, however, that labels can be lost (e.g., blown off, torn off), painted over, covered in dirt, rained on, become unreadable, etc., particularly when there are long-term isolations.

Ideally, the lock should prevent the isolation from being physically changed, especially for higher risk isolations. Use of a padlock provides a greater level of security than a plastic tag or wire loop, since the latter do not necessarily prevent the equipment from being operated. Where equipment is not designed to be fitted with a lock, e.g., a wheel operated valve, a chain and lock might be used, although this may not necessarily prevent the equipment from being operated. This issue should be considered as part of the risk assessment.

Where multiple trades are carrying out work, multi-hasp or lock box devices (Brady 2023) should be used where individual or trade padlocks are fitted (examples are shown in Figure 8-4). This ensures that de-isolation cannot take place until all of the locks have been removed.

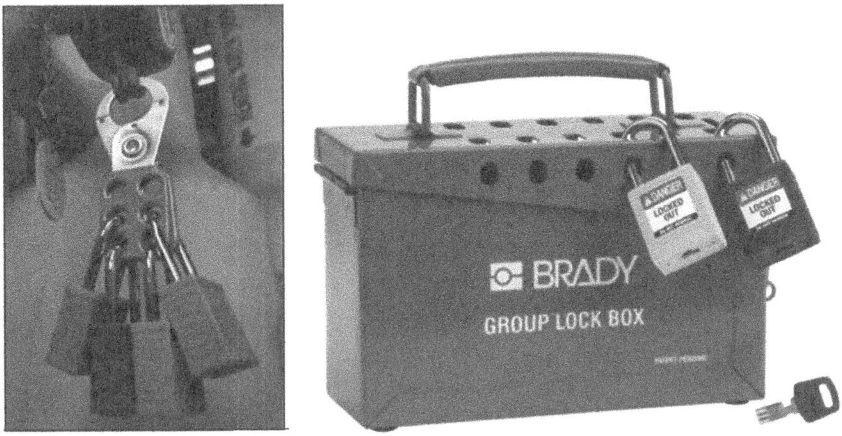

Figure 8-4 Multi-hasp and Lock Box

The best practice is to create a master document indicating the type of tag/label and its location as marked on a P&ID or electrical drawing. This master list can be used for removing all tags and returning the equipment to service.

8.5 DOCUMENTING ENERGY ISOLATION

With the exception of simple low risk isolations, the Energy Isolation Plan and activities should form part of the documentation pack that includes initial shutdown and isolation, through depressurization, decontamination, initial line break, testing and recommissioning. Workers must be informed and understand the contents of the documentation pack to ensure that they understand the risks, the mitigations (including isolations) that are in place, emergency plans and their role in ensuring that the work is conducted to plan. Typical documentation required is shown in Table 8-2.

Table 8-2 Typical Energy Isolation Documentation and Respective Usage

Document Type	Document Usage
P&IDs	Marked up with numbered isolation points, purge and vent points, break-in points
Isometrics/ schematic drawings	Same as P&IDs
Photographs	Labelled to clearly identify isolation points, purge and vent points, break-in points
Single-Line Electrical Drawings	Marked up with numbered isolation points, earth points
Equipment Drawings	Marked up with potential locations of trapped energy, isolation points
Procedures	Shutdown, depressurization, decontamination, purging, isolation, LOTO, first line break procedure, PPE requirements, potential remaining hazards, essential/non-essential personnel, barriers, notices
Checklists	Steps for isolation, decontamination, purging, LOTO, including dual sign-off if required by risk assessment
LOTO Master document	Showing all process, trade and individual locks and their location.
Cross reference to other work	Details of other work being carried out on the same system or the same area (Simultaneous Operations - SIMOPS) that may be using the same isolations or that may affect the work to be conducted
Safe Work Permit	Includes: work to be conducted, energy isolations, mitigations (PPE), residual hazards, work duration, authorized personnel, emergency procedures and communication and cross referencing all other documents in this table.

An example Safe Work Permit is provided in Appendix A.

8.6 CHEMICAL PROCESS ISOLATION

The responsibility for chemical energy isolation typically lies with the process operations team and their first consideration involves isolation from sources of pressure, temperature, fluid (liquid or gas), toxic, flammable, reactive materials, asphyxiants, and materials that may become unstable when they are dried out. The initial step to developing an isolation plan is to fully understand the unique hazard characteristics of the chemicals in the process. Relatively low concentrations of some chemicals in the air can create explosive atmospheres and even lower concentrations of the same chemicals may pose personal health hazards. The Safety Data Sheets provide information on the hazards of the materials as discussed in Section 3.3.1. Processes containing solids and liquids may create unique isolation and decontamination challenges, such as plugging, draining, drying, or neutralizing the chemicals. Thoughtful and thorough consideration of the process conditions, the chemical components, and the potential hazard consequences followed by an appropriate risk assessment will be required in order to develop a safe and appropriate isolation plan.

8.7 ELECTRICAL ISOLATION

There is a general awareness of the basic hazards involving electricity in the domestic environment. For example, not to use electrical devices around water, not to insert non-approved items into electrical sockets or use electrical equipment with frayed electrical cords, and only to use approved items in fuse boxes. In the chemical industry, the vast array of installed electrical systems creates challenging safety requirements and a multitude of additional risks, particularly when considering isolation for maintenance, that may include:

- Injury to personnel due to electrical shock or possible health issues due to electromagnetic radiation (e.g., from high voltage transformers)
- Damage to equipment due to short-circuit or overload and possible fire.
- Accidental starting of a motor or movement of an actuated valve, that could affect the isolation, or cause direct injury to personnel, or equipment.

Electrical systems will range from a few milliamps to many thousands of volts, single phase to three phase. Work, including isolation on high voltage systems (e.g., 1 kV or above) requires special attention and most regulators require Subject Matter Experts (SMEs) to be involved where there is work conducted on such systems. This may include the consideration of arc flash hazards and special PPE requirements.

Much of the primary process equipment, such as pumps, mixers, blowers, fans, centrifuges, blenders, agitators, and compressors are supplied at 220 - 440 volts; that is sufficient to cause a fatality if it comes into contact with a person. Understanding the source and function of the electrical power, its configuration, and isolation points is critical to performing safe and complete isolation. Facilities should have readily available electrical drawings to show the equipment and the isolation points. It is particularly important to be aware that some designated safety equipment or processes may have automatic back-up power supplies, such as an uninterruptible power supply (UPS), thus, *all* sources of electrical power that present a potential danger must be identified and isolated. Some electrical equipment may contain a battery or capacitor that remains energized even when disconnected. Proof testing of the isolations is required, to demonstrate that the equipment will not operate after being isolated. Many organizations use a Lock-out Tag-out Try-Out system for electrical isolations as further detailed in Section 9.6.

Case Study 8-1 — Electrical Low Voltage Isolation

Description

Modifications were being made to wiring from process equipment to the process control room. Electrical isolation of the 110V system was implemented at the electrical breaker boxes using approved isolation devices. Unfortunately, the electrician contacted a live low voltage wire when in the attic, causing loss of life.

Comments

The existing control wiring was low voltage but relatively confusing due to changes over the years. The work involved crawling in the attic area above the control room. The area was well lit, but very tight for moving around. The isolation plan for the electrical wiring project did not identify that some of the wiring remained powered, and that the working area was also near some live wiring. The isolation plan only addressed the equipment and wiring that was being modified.

Key Learning

Although the scope identified specific electrical isolation points, additional electrical sources and their risks were not considered. A more thorough review of active energy sources at or near the working area might have identified the risk and prevented the incident.

8.8 MECHANICAL, POTENTIAL AND NATURAL ENERGY SOURCES

Equipment that can fall or move due to the action of gravity, a spring, water flow, or wind must be considered as part of the energy isolation plan. Counterweights are an example that can cause movement of equipment even if other power sources such as electricity have been isolated. For example, in the oil fields one type of pumping device for the well is called a "horsehead" because of its physical appearance. This pump has a large cantilever weight assisting the motor in powering the rotation. Even after the electrical power to the motor has been isolated, the cantilever arm will continue to move. Another example is spring loaded isolation valves where the spring is designed to force the valve to fail to a safe position upon loss of instrument air or electricity. A further hazard is engulfment, which is a potential hazard when working in or beneath significant quantities of solids, such as grain, pellets, or powders.

Understanding the system design and how to fully isolate it from creating an unsafe action must be considered in the isolation plan that may include locking or tying components in place. Subject matter experts may be required to ensure the method used to immobilize the equipment is adequately designed.

When equipment is bolted to or used to support a load bearing structure, caution should be given when unbolting or disconnecting the equipment. The equipment, connected piping, or even the steel supports may spring or jump once disconnected. Potential consequences in these situations could range from a minor personnel injury to a major chemical release if the piping or equipment should break open. A system whereby individuals are located out of the line of fire may be appropriate under such circumstances.

Water or wind power may create a unique consideration for isolation. Even after the source of the wind or water flow have been turned off, the equipment may continue to rotate or spin. Additional mechanical isolation may be necessary to completely isolate and deactivate the motion of the equipment.

Photovoltaic panels require special consideration where solar energy generation may be zero when the isolation plan is considered but could start when the sun shines.

8.9 NUCLEAR RADIATION

In addition to electrical radiation as discussed in Section 3.3.4, radiation from nuclear material must not be overlooked as outlined in Section 3.3.7. Naturally Occurring Radioactive Material (NORM) can be present on oil and gas production equipment and can accumulate in certain areas, such as separator drums. Where NORM might accumulate, a radiological survey, conducted by a subject matter expert, should form part of the isolation/decontamination plan.

Nuclear level gauges are often used due to their high levels of accuracy and reliability. The nuclear sources must be isolated or removed before personnel can safely access the affected area (typically involving vessel entry). The isolation or removal of the radiation source must follow the method or procedure provided by the Original Equipment Manufacturer (OEM) or other subject matter expert. Certain countries and states may require a licensed technician to perform the isolation procedure. The timing or sequencing of performing the Isolation must be considered in the overall decontamination plan since the nuclear level device may be necessary, to measure when the tank or vessel is empty.

Other potential sources of nuclear radiation include X-ray inspection equipment that is often used during turnarounds. The use of such equipment must be the subject of its own risk assessment and would likely include barriers and signage to prevent personnel entering areas that may be subjected to radiation.

8.10 SUMMARY

This chapter provides an overview of the various types of energy isolation, including LOTO and discusses their relative security. It details some of the issues that can be encountered when specifying energy isolation, consideration of human factors, and the requirements of the associated documentation. More specific isolation requirements are discussed further in Chapter 9.

9 WRITTEN ENERGY ISOLATION PLANS (EIP)

An Energy Isolation Plan (EIP) should be a required component of a company standard or corporate procedure and forms part of a holistic approach to safely manage the entire process of preparation of equipment for maintenance and return to service. An EIP comprises a series of steps to be followed to ensure that all sources of energy that could cause harm to people, the facility, or the environment are removed before work is conducted on a piece of equipment or part of a process system. A LOTO procedure is the physical application of tags and locks to ensure that the Energy Isolations cannot be inadvertently removed and will typically either be part of an EIP, or a separate procedure referenced by the EIP. In the USA, OSHA standard 1910.147 provides regulatory standards (OSHA 2024) and further information is also available in the UK HSE guide HSG253 (HSE UK 2006). The authors of EIPs/LOTO should be trained and competent in the operation of the process to be isolated, as well as any previous risk assessments and isolation standards for the facility. Awareness and knowledge of local regulatory requirements for LOTO must be considered when creating the EIP. Workers who are using the EIP must understand and check the isolation points, using their own locks, as required. The various steps are summarized as follows:

- Document objective(s) of the EIP.
- Identify the potential hazards and consequences.
- Identify Energy Sources.
- Identify Isolation Points.
- Shut down the process using standard operating procedures, or temporary procedures that may have been developed for the specific activity to be conducted.
- Isolate Energy Sources.
- Lock valves and switches and other sources of energy in the safe position.
- Tag valves and switches.
- Check to be certain that energy is dissipated.
- Monitor the energy isolations regularly.

9.1 DE-ISOLATION, TESTING, AND RESTART CONSIDERATIONS

The EIPs and any associated procedures should also include consideration of return to service, with respect to de-isolation and recommissioning, to ensure each step can be conducted safely. Examples of issues associated with de-isolation that should be considered at the isolation planning stage may include:

- Should a process stream be isolated via a blanking plate or blind if there is no provision to check the integrity of the upstream isolation prior to removal of the blank or blind?
- Can purging be accomplished prior to removal of the isolation? This may include purging, passivation, or catalyst reduction stages.
- Can pipe spools be reconnected, where piping has been unsupported and may have moved?
- Should a disturbed joints register be used, where any disturbed joints, including small bore fittings are tagged, recorded, and tracked throughout the maintenance period, up to and including testing and recommissioning? Further details on the use of a disturbed joints register are provided by the Energy Institute's publication: *Guidance on Safe Plant Reinstatement* (EI 2023).
- Does safely re-commissioning the isolated process require that utilities, services, and process chemicals be returned to service in a designated sequence?

9.2 RISK LEVEL

Chapter 8 provides an overview of Energy Isolation and some examples of different options and types of isolations and devices that are used to ensure safe isolation is maintained throughout the work activity. Simple isolations, where the potential consequences are low, such as a low pressure water system where the worker is in sight of the isolation valve, could be safely addressed using a Work Permit. For more complex or potentially hazardous isolations, a documented EIP is required. Examples include:

- Work on equipment containing pressurized, toxic, flammable, hot, cold, radioactive, or biohazard material.
- Work requiring isolation of multiple energy sources.
- Isolations required by multiple trades or organizations.
- Isolations that are required for multiple tasks on a common system.
- Isolations that affect more than one process or production unit.

- Isolations for confined space entry.
- Isolations to prevent physical injury by unplanned startup of equipment.

The last point is well illustrated by a historic incident that occurred in the UK in 1972 and was referenced by an Imperial Chemical Industries Safety Newsletter in August 1972 (ICI 1972) as detailed in Case Study 9-1.

Case Study 9-1 — Lack of Energy Isolation

Description

"A fairground engineer escaped with minor back and leg injuries after being fired from a fairground cannon at Loughborough yesterday. He was half-way down the barrel of the 16 ft. cannon adjusting the mechanism when it accidentally went off."

Comments

Presumably the energy was stored in a large spring and the trigger mechanism, or the spring, was not physically disabled.

Key Learning

- Energy Isolation must include all sources of energy including instrumentation and control systems (and trigger mechanisms!)

The EIP is prepared following a risk assessment and prior to commencement of the work and must specify details of all isolations that are required from each source of energy, and an appropriate level of approval based on the magnitude of the risk must be included.

Some sites consider the adequacy of energy isolation in the following terms:

- A single form of energy isolation may be adequate for maintenance work on low pressure water lines, or for work on a 120-volt electrical system.
- Double isolation with intermediate venting (double block and bleed arrangement), where there is flammable or toxic material present.
- Double isolation for electrical work on high voltage systems – e.g., rack out contactors and remove fuses (and possibly provide an earth or grounding).

- Single valve isolation of dangerous materials can only be used for the purpose of fitting a higher form of isolation, such as fitting a line blind or physical disconnection. Full risk assessments are required, including a high level of management approval. PPE requirements must assume that the primary isolation is not functioning.
- Where the consequences of error are major, for example for vessel entry, two diverse forms of isolation are recommended. For piping this may involve a valve and physical disconnection. For electrical devices, fuses should be removed, and drive or electrical leads disconnected.

In situations where it is not possible to achieve the required integrity of energy isolation, it may be necessary to review the isolation envelope to include other parts of the process that are not being worked on but can provide improved isolation for the work area (see Case Study 9-2, which revealed inadequate isolation prior to a confined space cleaning process). Alternatively, other layers of protection (barriers) should be put in place and specified on the EIP. However, where a company standard cannot be followed, this should be escalated to a senior level for review and authorization. Once an EIP is approved, any changes subsequently made will require careful consideration under the Management of Change (MOC) process.

Case Study 9-2 — Inadequate Isolation

Description

A reactor at a facility in Mexico required isolation after every batch, for confined space entry for cleaning purposes. A safety audit was being conducted while the vessel was being cleaned internally, and the auditor observed that the isolation valves were closed but not locked. Furthermore, the valves were also located on three different levels and were out of sight of the person on standby.

Comments

Any of the ball valves could have easily been opened without the standby man knowing that was done, potentially leading to a major incident or even a fatality. The auditor stopped the work and informed the plant manager and operations manager.

Key Learning

- Energy isolation plans for confined space entry should be thoroughly risk assessed to ensure they are adequate.
- Periodic audits should be conducted to ensure Energy Isolation Plans are (a) adequate and (b) carried out correctly.
- Stop Work Authority: anyone on site should have the authority to request work to be stopped if there is a concern about safety.

9.3 DOCUMENTATION - SAFE WORK PERMIT AND EIP

The EIP must formally connect to the work permit, either by a reference system or by being physically attached to the permit. The EIP should include sketches or marked-up P&IDs/isometrics/electrical drawings with a unique reference number for each isolation. That reference should also be listed in a table that allows workers to identify all of the isolations and the jobs for which each isolation has been made. Depending on the nature of the work, other documentation that may require referencing via the work permit includes hot work permits, confined space entry permits, excavation permits, etc. The supervisors for each trade (and ideally all workers) must understand the contents of all documents, including associated symbols as detailed in Section 9.4. This allows the workers to put their own lock on the isolation(s), further protecting them from someone in another trade accidentally compromising the isolation while the work is still in progress.

Case Study 9-3 — Filter Cartridges Change Out

Description

A practice for a chemical unit was replacing filter cartridges on a regular basis. There were two banks of filters in parallel. The replacement practice was conducted by operating personnel of the unit following a written change procedure. A junior operator who had only changed the filters once before was tasked with replacing the filter cartridges. Labelling had degraded, which created confusion for the junior operator, who isolated the wrong set of filters, which subsequently led to the opening of a drain valve on a live filter. The operator's feet and lower legs were sprayed with a harmful chemical, causing skin burns, resulting in a lost time injury.

Comments

Replacing filter cartridges was considered a routine work procedure, but the consequences associated with opening the wrong set of filters were not adequately communicated to junior personnel. Labelling of isolation valves and points had degraded. A formal LOTO had not been considered necessary for this routine task.

Key Learning

- Routine tasks involving opening equipment that handles harmful chemicals should be conducted using an Energy isolation Plan.
- A Work Permit in addition to the Standard Operating Procedures may have prevented this incident.
- Labelling of process equipment isolation points and vent/drain valves must be maintained and easily readable.

9.4 PIPING AND INSTRUMENT DIAGRAM (P&ID) SYMBOLS FOR PROCESS ISOLATION

Some typical symbols that are used on P&IDs associated with process isolation are provided in Figure 9-1. Personnel conducting the EIP and workers using the EIP should be familiar with these, or locally used symbols.

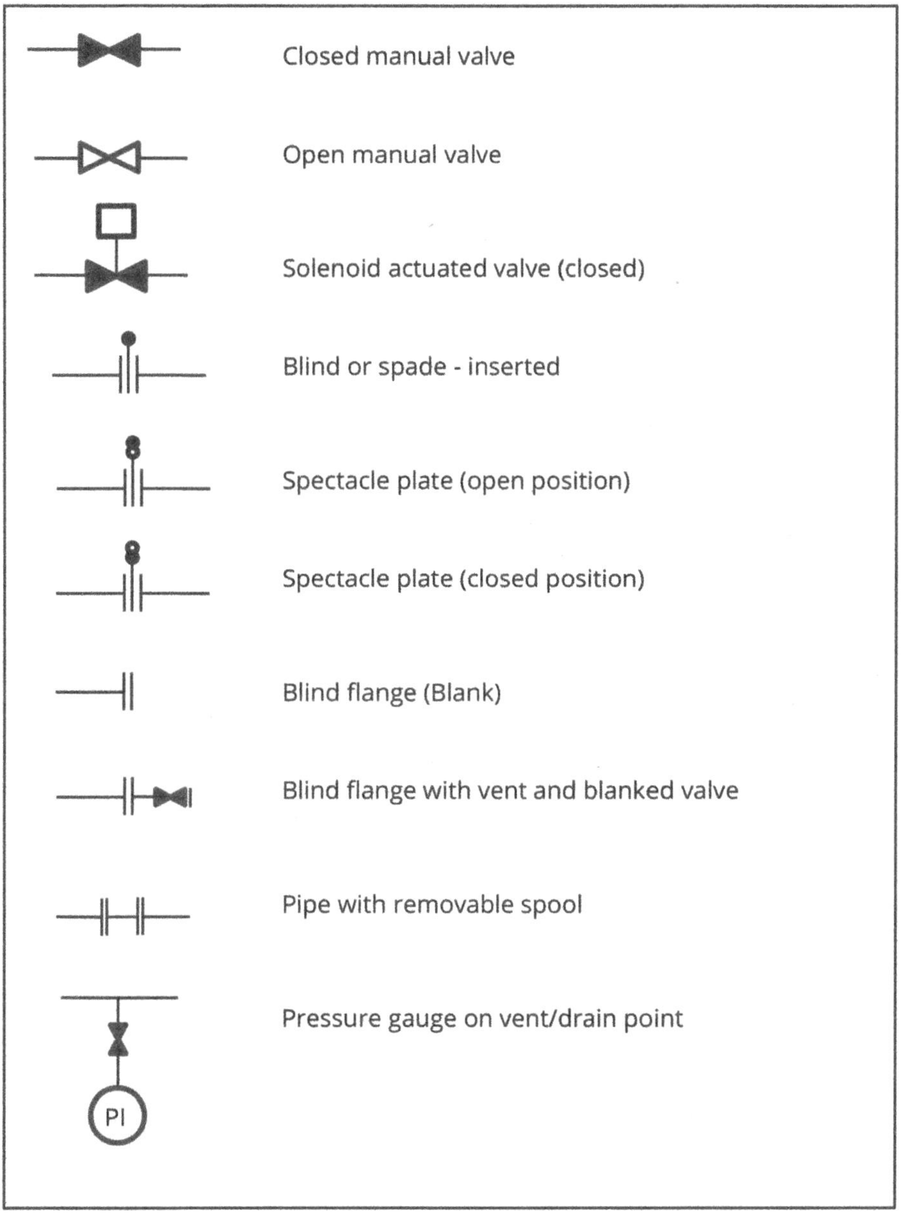

Figure 9-1 Key to Typical P&ID Symbols for Energy Isolation

9.5 FORMS AND TYPES OF ENERGY ISOLATION

As stated previously, different forms and degrees of energy isolation may take place at various stages of isolation and decontamination. For example, single process isolation may be used ahead of closure of a spectacle plate or removable of a spool and fitting of a blind flange, which then provides positive isolation. Ideally, a vent/drain point should be available, onto which a pressure gauge can be fitted to check the integrity of the single isolation valve (proved isolation). The pressure gauge would be checked first and if it reads zero, a valve could be carefully opened to atmosphere as a double-check. Opening of the valve should be considered a first line break (See 9.9 and the specified PPE should assume that the first valve is leaking by (passing). If no vent/drain valve is fitted, opening the downstream equipment should be treated as a first line break.

For electrical isolation, opening and locking an isolator would be carried out ahead of the removal of fuses and potential grounding. Further details on electrical isolation are provided in Section 10.2 and on the UK HSE website (HSE UK online). For isolating a source of potential energy, such as a heavy weight that could fall, a temporary sling might be used before something more robust, such as a structural scaffold, can be fitted.

If the line to be worked on is downstream of a pump, the electrical isolation of the pump might provide a secondary form of isolation, although electrical isolation of a pump on its own would not be adequate, especially if it is a centrifugal type.

Various types of process isolations in order of increasing integrity are shown in Figure 9-2. Details of other forms of energy isolation are provided in Chapter 10.

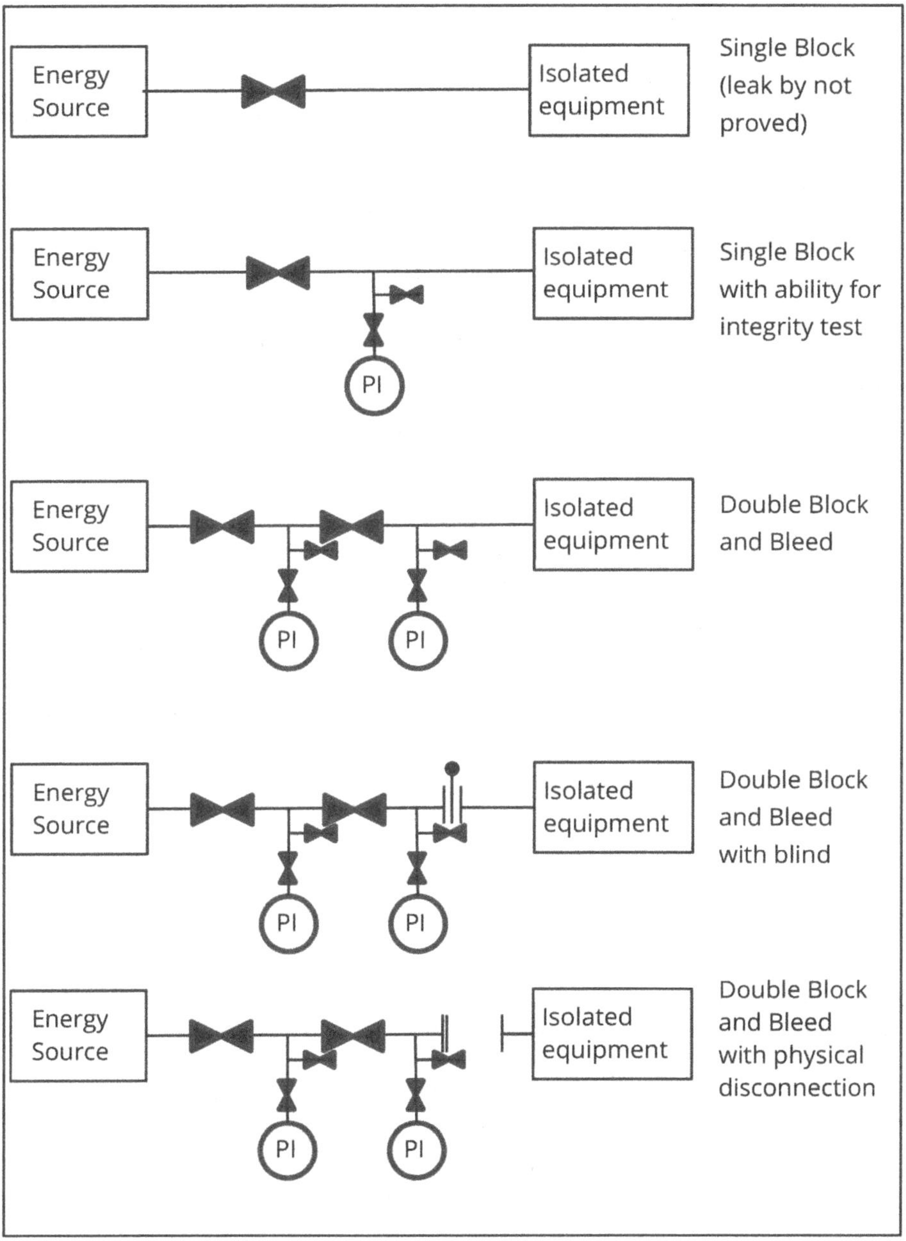

Figure 9-2 Process Isolation with Increasing Integrity of Isolation

9.6 LOCK - TAG - TRY

In addition to Lock-out Tag-out for electrical isolation, many companies include a "try-out" step ("Bump start") as part of the Energy Isolation Plan/LOTO. This involves a further check, where an attempt is made to start the equipment, as a final step to prove the electrical isolation. This can be an effective step although it must be used with caution. There have been incidents connected to the Try-Out stage in which the electrical starters are latching types (the circuit remains connected unless a Stop button is pressed). This has led to equipment starting up when a start signal is inadvertently sent to the equipment. A further stage—"Reset"—should be added to the "Try-Out" step.

9.7 FINAL VERIFICATION BEFORE RELEASE TO WORK

Once the energy isolation has been completed, this must be verified and validated in the field by Operations and Maintenance personnel "walking the line" and checking against the original EIP. This is a basic expectation of Conduct of Operations, sometimes called "Operational Discipline", to conduct a deliberate and structured verification that systems are energy-free, including checking bleed points for evidence of liquids or pressure, prior to handover for maintenance. For higher risk isolations, this should be carried out by an authorized, competent, and independent person. Particular attention should be given to any changes to the isolation plan, such as using a different valve when the original one was found to be seized or letting by (passing), as illustrated in Case Study 9-4, which involved inadequate isolation and a flash fire (CSB 2017a). If the change is significant, further risk assessment may be required. Further details are provided in Section 10.7.

Case Study 9-4 — Flash Fire Caused by Inadequate Isolation

Description

In November 2015, workers were isolating and de-inventorying part of a process unit at refinery in Wilmington, Delaware, to replace a 20 ft (7 m) section of pipe that had thinned. The pipe normally carried caustic soda, and the original plan was to isolate the section of line from the two feed sources and to drain it down. However, it was found that the single isolation valve from the caustic system was leaking by (passing) and operators had to develop a new plan that involved a significantly larger and more complex isolation. This included closing a valve to isolate the caustic system from a Depropanizer system and displacing the contents of the caustic system with water and draining it via a 3" valve to the oily water sewer.

As the system was being drained of water for the second time, there was a sudden release of flammable vapors that ignited, resulting in a flash fire, causing burn injuries to the operator. There were also fires from several other open drain valves on the system and the fire burned for about an hour before the onsite response team extinguished it by isolating the drain valves.

Comments

The investigation found that the isolation valve to the depropanizer column was also leaking by (passing), allowing hydrocarbons to pass back into the caustic system, and flowing out with the water as the system was being drained down. The probable source of ignition was a nearby furnace.

Post-incident analysis showed that there were three other valves between the caustic system and the depropanizer that remained open for the maintenance preparation activity.

Although the facility had a LOTO procedure for marking and tagging energy isolation, this did not include the preparation stages of proving isolations, de-inventorying, and washing/draining.

The facility had developed some procedures for taking specific equipment out of service for maintenance, but did not include the caustic system. The procedures for this specific activity were developed relatively informally.

Key Learning

- Single isolation, especially when it cannot be proven, is inadequate for isolation of hazardous energy sources.
- Double block and bleed arrangements allow the integrity of process isolations to be proven.
- Preplanning including risk assessment is required prior to taking equipment offline and preparing it for maintenance.
- A change in isolation plans requires the risk assessment to be redone.

  Isolations should be monitored to ensure their continued integrity.

9.8 FIRST LINE BREAK

Where a valve or other equipment that has contained any form of energy is first opened to atmosphere, this must be considered the "First Line Break". While there may have been some ability to check the integrity of the isolation, it is still possible for energy to be present in the form of residual fluids, trapped pressure (particularly in valves and non-return valves), or pockets of vapors. Many organizations have documented procedures for first line breaks that may include:

- Positive line identification programs where the line is clearly tagged or marked, to ensure the correct break location.
- Specified PPE requirements. This should assume that the worker could come into contact with the source of energy, until the isolation has been proven.
- Presence of a standby person with a radio to call emergency services, if required).
- Provision of spill containment. This could either be via a drain line or a suitable container, positioned beneath flanges or drain points.
- Use of appropriate gas monitors that have been tested.
- Keeping all other personnel away from the area until the First Line Break has been successfully completed.
- Using signs and barriers to ensure workers do not accidentally encroach into the potential danger area.
- Techniques for staying "outside the line of fire". Pipe fitters should be trained to carefully loosen the bolts, in small increments at a time, on the far side of the flange. If there is trapped energy, it should thus be directed away from the worker.

9.9 MONITORING ISOLATIONS

Each isolation device and tag must remain in place until all activities that require that particular isolation are completed and signed off. Some isolations may be required longer than others and, on occasions mistakes are made, such as isolation locks/tags being inadvertently removed, which may compromise the integrity of the isolation. These types of issues can be managed by periodic verification of the isolations, using the documentation provided when they were first carried out, and by Walking the Line. This verification can be conducted at all levels within an organization and should be formalized for the higher risk/longer term isolations.

The monitoring should include:

- Reviewing the documentation (permit to work, EIP, marked-up P&ID).
- Checking the visibility and condition of labelling and locks of isolations in the field.
- Ensuring any specified equipment such as barriers, gas detectors, and signs are in place and functioning.
- Identifying any modifications, either to the isolations, locking systems, or where vent/purging lines might work in tandem to compromise isolations.

Case Study 9-5 — Safety Relief Valve LOTO Compromised

Description

An LPG sphere was fitted with a safety relief valve and a manual isolation valve was car-sealed open, supported by a Lock-out Tag-Out (LOTO) system. During a routine housekeeping inspection around the ground area of the sphere, a plastic tag was observed half buried in the soil. It was not possible to identify the source of the tag, but a visual check at the top of the sphere revealed that the tag on the relief valve inlet was missing.

Comments

The plastic components of the tagging system had degraded over several years and eventually failed. Some LOTO equipment is not intended to be used for a long period of time and can degrade or fade due to sunlight, wind etc. The system did not use locks/chains and it would have been therefore relatively easy for someone to inadvertently close the isolation valve to the relief valve, rendering it non-functional in the event of an emergency.

Key Learning

- Car Sealing of valves and electrical equipment can be used as part of an EIP to isolate a plant for maintenance. However, for higher risk isolations, physical locking of valves, including locks and chains are much more secure and less likely to fail or otherwise be compromised.
- Periodic audits should be conducted to ensure required isolations remain in place.

9.10 SUMMARY

This chapter provides guidance of the content of written EIPs and how they are linked with LOTO, Work Permits and other documentation requirements. Some of the different forms of energy isolation are discussed, along with their respective level of risk, procedures for first line break, and ongoing monitoring of isolations.

10 SPECIFIC ISOLATION REQUIREMENTS

This section discusses specific details of energy isolation. As with all other isolations, they should be clearly documented to ensure that they have been correctly performed and that the de-isolation can be conducted safely and effectively.

10.1 PROCESS/INSTRUMENT ISOLATION

Depending on the level of risk, process isolation can range from a single valve to a double valve, double block and bleed, blinding, or ultimately physically disconnecting/blanking the pipe or equipment. Further details are provided in Section 9.6. For automated process valves controlled by the supply of air, hydraulic fluid, or an electric motor, it may be appropriate to also disconnect those sources to prevent inadvertent operation.

Standalone isolation of instruments is often required for routine calibration, repair, or replacement. Due to variable design features and installation configurations, instrument isolation requirements often are challenging and must be carefully considered to avoid spills or exposures. For example, the installation may include a single block valve, block and bleed valves, or no isolation valves at all; the last of which would require a detailed risk assessment, possibly at a high management level, with mitigation actions to accompany the first line break. When no isolation valves are present, the process equipment will most likely need to be fully isolated and decontaminated under careful supervision.

A detailed risk analysis is essential where pathways to pressure relief devices are obstructed by process isolation since equipment can quickly become over-pressurized during purging or pressure/leak testing, as illustrated in Case Study 10-1, where missed steps in the isolation process led to multiple lives lost, injuries, and plant damage at a Phillips 66 facility (HSE UK 1989). Should the pressure relief path be compromised, an alternate relief path must be available, or the decontamination plan be fully re-evaluated. More specialized process isolation methods are detailed in Section 10.6.

Case Study 10-1 — Vapor Cloud Explosion at Phillips 66 Facility

Description

A specialist contractor was carrying out a maintenance activity on a High Density Polyethylene (HDPE) process. This involved unblocking the settling legs in the reactor loop, where particles of HDPE from the slurry in the reaction formed a blockage. As the work was being undertaken, one of the legs in the reactor loop suddenly cleared a blockage, releasing a large cloud of flammable vapor. The cloud ignited resulting in a large vapor cloud explosion that damaged other process equipment and led to further losses of containment, explosions, and fires. The incident resulted in 23 fatalities, 130-300 people injured and extensive damage to plant.

Comments

The process isolation for this activity was provided by a single, 8" pneumatically operated ball valve. Although the Phillips Pasadena facility's local procedure required the air line to the actuator to be removed, that was not done. It had also been incorrectly connected, such that the valve was open when it should have been closed. Once the blockage cleared, there was nothing to stop the reactor loop from depressurizing to the open end of the pipe.

The Phillips corporate standard isolation procedure required double isolation or the use of a blind flange for this type of activity, but that was not adopted at this particular facility.

Key Learning

- Double isolation should be the minimum standard for high-risk isolations.
- Energy supply isolation to valve actuators (pneumatic, hydraulic, or electrical) must be included as part of the isolation plan.

10.2 ELECTRICAL ISOLATION

High voltage electrical isolation, typically above 1000 volts, (definitions may vary) requires special consideration and multiple steps, performed only by suitably qualified personnel. Isolation of electrical equipment at lower voltages can be performed by turning off an isolator in combination with a LOTO scheme, or removal of a fuse, disconnection of cabling, etc.

Some electrical equipment should also be grounded for additional protection. The degree of electrical isolation required will be a function of the level of risk.

For example, if work is to be conducted on an agitator externally to a vessel, a lower degree of electrical isolation may be appropriate compared with the requirement for vessel entry. Regardless of the type of electrical equipment, however, the principle of LOTO and Try-Out ("Bump-start") is a good standard practice (see Section 9.6), in which an attempt is made to start the equipment under carefully controlled conditions as an additional level of safety before determining that the equipment is in fact isolated. As discussed in Section 9.7, if there is a "reset", this should be used following the Try-Out in case the equipment potentially remains live after the Try-Out.

An example of an electrical isolation panel from a training center is provided in Figure 10-1, showing isolation locks and green start/red stop buttons.[4]

Figure 10-1 Electrical Switchbox Used for Training

[4] Photo courtesy of Teck Industries training center

While the applications may differ, the 7 steps for safe electrical isolation are similar for both low voltage and for high voltage. The process is as follows (Electrical Safety UK online):

1. Identify suitable point(s) for isolation.
2. Carry out the safe isolation process.
3. Secure the point of isolation (attach locks and tags).
4. Prove the circuit is dead.
5. Apply safety earths [grounding].
6. Place danger notices warning that adjacent electrical equipment may be live.
7. Issue an electrical permit to work.

Electrical isolation must be conducted by personnel who have received appropriate training. For example, the UK's The Electricity at Work Regulations 1989 (HSE UK 2015) requires that those engaged in electrical activities have the knowledge and experience to undertake work safely. Also, the NFPA 70 National Electrical Code and NFPA 70E, Standard for Electrical Safety in the Workplace, reference required training and the use of appropriate PPE (NFPA 2024). Employers must demonstrate that their staff have been deemed competent to undertake safe electrical isolation tasks.

10.3 MECHANICAL AND POTENTIAL ENERGY AND NATURAL ENERGY ISOLATION

Sources of energy can sometimes be hidden, such as the "spring" that can occur in a pipeline, whereby it suddenly moves and can strike a person or piece of equipment when flange bolts are removed. Gravity also should be considered in circumstances where equipment can suddenly fall or slide when it is disconnected. Under such circumstances it may be appropriate to fit a physical restraint such as a wired sling. Expert advice should be sought, where required, to assess the type and strength of such restraints.

At the time of writing, there were major issues concerning the mangled steel bridge components from the Baltimore, Maryland, bridge collapse that have the potential to move/spring significantly due to trapped energy, as the bridge debris is cut up for removal and disposal. Controlled demolition using explosives *is considered the safest and most effective way to remove steel under a high level of pressure and tension* (Newsday 2024).

Large rotating equipment such as fin-fans may start rotating when wind speed picks up. On equipment where locks are not fitted it may be appropriate to fix temporary clamps. For example, when working on cooling tower fans, scaffolding planks could be locked in place to prevent rotation. Again, expert advice should be sought, and the method should be subjected to risk assessment.

The renewable energy sector has introduced new types of equipment that may require special consideration for energy isolation during an unexpected change in the conditions, such as:

- Increased wind levels for wind turbines
- Increased solar power for photovoltaic cells (solar farm)
- Changes in demand (energy storage schemes)

10.4 NUCLEAR RADIATION

On regulated sites in the nuclear industry, radiological protection engineers and other specialists are responsible for conducting risk assessments and establishing procedures for isolating and moving or removing nuclear material. Safety, security, and lifecycle aspects of nuclear materials are usually controlled by regulators such as the Nuclear Regulatory Commission (NRC) in the USA and the Office for Nuclear Radiation (ONR) in the UK. Environmental aspects on lower level waste material may fall under a separate regulator such as the Environmental Protection Agency (EPA) in the USA or the Environment Agency (EA)/Scottish Environment Protection Agency (SEPA) in the UK.

In the petrochemical industry, specific measures should be taken to avoid exposure to NORM, Natural Occurring Radioactive Materials. NORM, especially radon and radium, are present during oil and gas extraction and are mobilized by the water present in the wells. The NORM can precipitate out of solution, with sulphate and carbonate deposits as scale and sludge in pipes and related equipment. Exposure concentrations are typically low, but precautions are recommended when handling scale and sludge or conducting maintenance on equipment coated with such material. Safe handling practices include personal monitoring and good hygiene practices.

Nuclear sources used in the process and other industries, for example, as used in radioactive level gauges, should either be isolated by closing (and locking) a lead barrier, or be physically removed and relocated to a purpose-built, lead-lined container. This must be conducted using a procedure that has been developed specifically for this purpose.

10.5 BOUNDARY ISOLATION

Boundary isolation involves large sections of a plant, such as a process unit, being shut down and isolated from the rest of an operating plant or facility for an extended period of time, such as during a maintenance turnaround. Special procedures developed for boundary isolations should be at least equal to, or exceed, normal isolation standards for the site. As a minimum this should involve blind flanges that are rated for maximum pressure, and ideally physical disconnection of feeds, product lines, and services.

It is likely that certain services or pipe bridges that pass through or are close to the unit may remain live. Job risk assessments must consider these aspects and workers must be made aware of any systems that are still live in their work areas.

10.6 SPECIALIZED ISOLATION TECHNIQUES

In the event that process isolation cannot be achieved through other means, highly specialized methods of isolation are available, although these are rarely used beyond low risk operations, such as those handling water. They must be:

- Considered as a last resort where mechanical means of isolation are not available.
- Subjected to a thorough risk assessment.
- Conducted by specialist, trained contractors.

10.6.1 Freezing

This involves freezing the fluid in a pipe and surrounding a section of the line with a cold medium, such as liquid nitrogen. Where nitrogen is used, it is allowed to boil off, which carries a significant asphyxiation risk, which must be considered as part of the overall risk assessment. The materials of construction of the pipe and associated fittings must also be considered, as their premature failure could result in loss of the isolation.

10.6.2 Pipe Plugs or Pigs

A pipe plug (or pig) is inserted into an open end of the line and a sealing system is provided against the inner surface of the pipe wall. This technique requires a separate primary means of isolation, such as stopping the flow in a pipeline, and can provide short-term isolation, to allow flow to be restarted temporarily. The design and materials of construction must be compatible with the process fluids and associated pressure.

10.6.3 Inflatable Bags

An inflatable bag is inserted through a blind flange or an open hole in the pipe and is then inflated to create a temporary seal. As for pipe plugs, this technique also requires a separate primary means of isolation, such as stopping the flow in a pipeline, but it can provide short-term isolation, to allow the flow to be restarted temporarily when needed. The design and materials of construction must be compatible with the process fluids and associated pressure. Careful monitoring of the bag pressure throughout the period of isolation is essential.

10.6.4 Foam Bags

Foam bags are similar to inflatable bags, but an expanding foam is injected into a semi-porous bag to create a more sustainable seal. This type of seal is difficult to remove and would normally be only used for permanent isolation; however, it would still require a form of secondary isolation such as a blind flange.

10.6.5 Squeeze-off

A mechanical or hydraulic vice or clamp is used to pinch or squeeze the pipe together to form a seal. This technique can be used for very low risk isolations.

10.7 FIELD VERIFICATION AND CHANGES TO ISOLATION

Whatever isolation method is used, field verification must be carried out before the work permit is released.

If work is being conducted over several days or weeks, it may be appropriate to recheck isolations and to reissue the work permits on a daily or shift-change basis.

Any changes to the isolation plan that are made in the field during the isolation and decontamination process and at other times must be considered, documented under the Management of Change Procedure, and communicated to all who are responsible for performing the work.

10.8 SUMMARY

This chapter provides details regarding specific and specialized energy isolation requirements and methods. Each isolation must be subjected to a full risk assessment that balances the hazards and risks with the reliability of the technique used.

11 REVERSING THE ISOLATION PLAN

This chapter discusses reversal of the Energy Isolation Plan (EIP). Recommissioning, Pre-Startup Safety Review (PSSR), and startup are addressed in Chapters 12 - 13 - 14. As discussed in earlier chapters, checklists are essential for isolation and decontamination. However, de-isolation is not necessarily a mirror of the original isolation plan—it requires careful planning to avoid damage to equipment, injury to personnel, or environmental damage.

11.1 JOB COMPLETION AND TRANSFER OF EQUIPMENT OWNERSHIP

Prior to starting de-isolation, it is essential to check that all planned work has been completed, inspected, and signed off by the responsible supervisor. The Operations team must physically check that the equipment is ready to de-isolate. Work Permits should be returned to the permit issuer and closed out before the equipment can be considered "handed back" to operations. Locks and tags will require removal by each person who applied a lock and signed a tag. If that person is not available, plant site safety practices may allow for removal of locks and tags after management review and approval.

11.2 DE-ISOLATION PROCEDURE

More complex and higher risk de-isolations require a documented procedure. This procedure must provide instructions specifying the order in which de-isolation shall occur/who is responsible for each step, and it must be associated with a checklist and a Work Permit. The isolation points should have already been clearly marked (See Chapter 8). Equipment owner responsibilities must be defined, and the owner must examine the equipment to ensure it can be put back into service according to the procedures detailed in the plan. The energy isolation device may include a multi-hasp lock or a lock box to ensure that the equipment cannot be de-isolated until the craft padlocks have been removed. This ensures that all work has been completed and no one is working on the equipment. As detailed in Case Study 11-1 and shown in Figure 11-1 (Cullen 1990; EI March 2023), the Piper Alpha offshore accident is a tragic reminder that significant consequences can result from inadequately isolating equipment.

Case Study 11-1 — Piper Alpha

Description

The Piper Alpha platform offshore drilling rig was located in the British North Sea, 110 miles northeast of Aberdeen, Scotland, and commenced operations in 1976. It initially operated on liquids only and the gas was flared off, but gas facilities were subsequently modified to recover gas via a compression station, flash drum, and l-out (KO) drum. The condensate was reinjected into the main oil line.

At about 2200 hrs on July 6, 1988, there was an explosion that led to a large fire that spread rapidly, followed by a number of small explosions and then a much larger explosion, at about 2250 hrs, when a gas main ruptured, destroying firewalls. Structural collapse began shortly afterwards, followed by further explosions. By about 0045 hrs, most of the platform had collapsed. The incident led to 167 fatalities—the highest number of casualties in any offshore accident.

Figure 11-1 Remains of Piper Alpha on the Morning Following the Explosions and Fires

Comments

The platform included two condensate injection pumps, one online and one on standby. To change the operational status, manual operation of valves and pump starts was required. Each pump was fitted with a pressure safety valve (PSV) located one floor above the pumps. On the day shift, prior to the incident, the PSV on the "A" pump had been removed for scheduled recertification but had not yet been replaced. The open ends where the relief valve was located were fitted with a blind flange, but the joints had not been tested up to system pressure, as they were not intended to be operated until after the relief valve was replaced.

At about 2145-2150 hours, the "B" condensate pump tripped out, and operators wanted to start the "A" pump to prevent a high level in the KO drum that would have tripped the gas plant. It was known that the "A" pump was out for maintenance and a work permit for preventive maintenance (PM) on an instrument associated with the "A" pump was located. The shift did not know that the "A" pump PSV had also been removed under a separate permit. Evidence suggests that the pump was electrically de-isolated and put online. Various gas alarms sounded, and workers heard a high pitched noise before the first explosion. The most probable immediate cause of the event was considered to be a condensate leak from the blind flange on the "A" pump PSV location, which had not been leak tested at system pressure.

The platform operated a Permit to Work (PTW) system that included details of the work to be done, process and electrical isolations that were carried out, any extra precautions, and required PPE.

Key Learning(s)

Many factors contributed to the cause and the extent of this incident as detailed in the 500-page Cullen report. In relation to the subject matter of this book, the report was critical of the procedures for isolation for maintenance including:

- There was no system specified for tagging or locking off process isolation valves.
- There was no cross-referencing between permits involving multiple tasks on the same equipment.
- There were no reviews of permits at shift handover.
- Suspended PTWs were not kept in the control room.
- There was inadequate training on the PTW system.
- There was a lack of consistency in using the PTW system between shifts.
- There were no periodic audits of the PTW system.

11.3 PROCESS DE-ISOLATION

A typical difficulty when de-isolating equipment is discovering that a line blind or blind flange provides the final isolation, but there is no way to check whether the primary isolation may have failed through the outage period. This is illustrated in Figure 11-2, where it is not possible to prove that the valves have leaked by/are passing material and pressurized the space between the valve and the blind flange. Ideally, the amount of material would be small, but this de-isolation requires careful planning that may include applying physical barriers; restricting the area from non-essential personnel; wearing appropriate PPE; and staying out of the line of fire (e.g., by first loosening bolts on the far side of the flange).

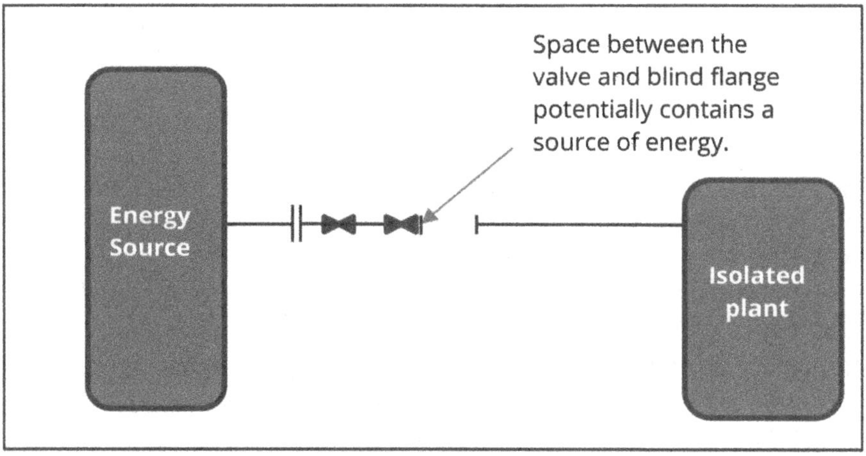

Figure 11-2 De-isolation with Unproven Absence of Energy Source

A safer and easier option for de-isolation can be conducted when the isolation has been performed via a double block and bleed arrangement. This allows the pressure between the isolation valves to be checked before the bolts on the blind flange are loosened. A recently calibrated pressure gauge can indicate whether the interstitial space between the valves has become pressurized so that venting to a safe location can be conducted. Care is still required, especially if there is potential for the bleed/purge point to be blocked, particularly where it is facing downward as shown in Figure 11-3.

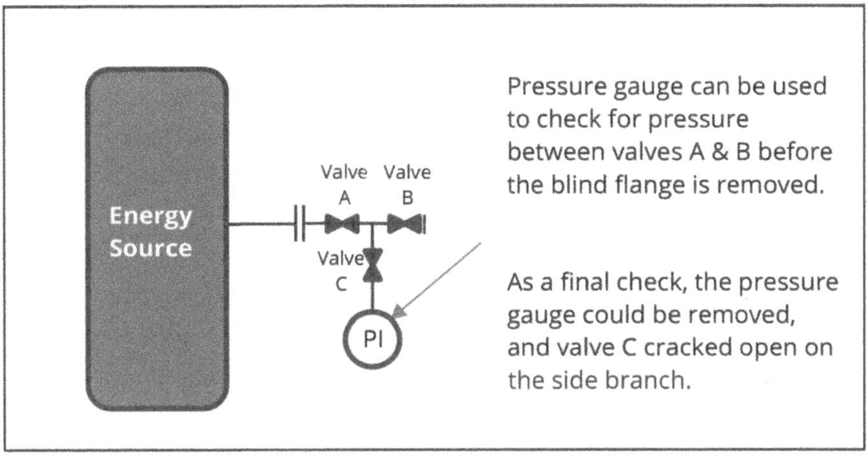

Figure 11-3 De-isolation with Ability to Prove Absence of Energy Source

Alternatively, a short pipe and bleed valve could be added to the blind flange to allow a pressure gauge to be attached, or the valve can be carefully cracked open and tested for any residual energy, as shown in Figure 11-4.

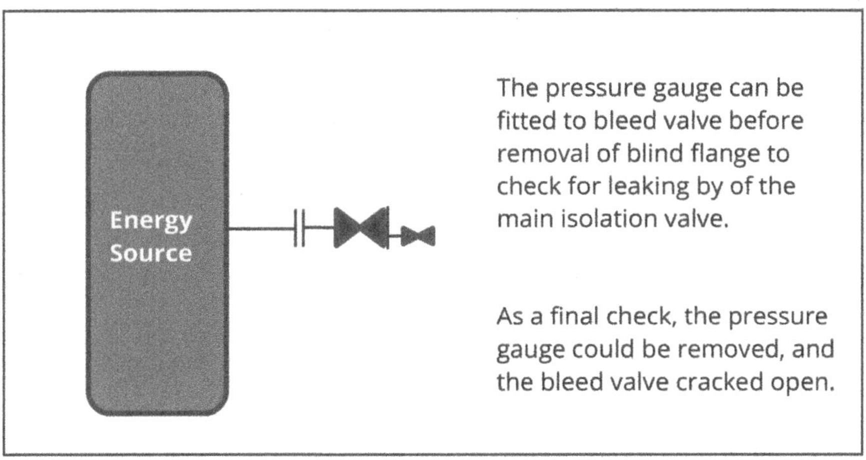

Figure 11-4 De-isolation with Bleed Valve Allowing to Check for Absence of Energy Source

11.4 VENTS AND DRAINS

The procedures and checklists used for de-isolation must include systematic checks for all flanges, drains, and vents that were opened as part of the original isolation plan. Some isolations or vent and drain lines may remain in place until the next stage of recommissioning is complete. For processes handling toxic or flammable chemicals, it may be necessary for the temporary tubing, hoses, and piping used for decontamination and purging to remain in place, connected to vent and drain valves, until drying and inerting during re-startup are completed. Following the procedures and documented steps to remove all temporary tubing and hoses, and securing the vents and drain ports, as outlined in the plan, are critical to preventing leaks when re-introducing chemicals to the process.

11.5 DE-ISOLATION OF OTHER SOURCES OF ENERGY

The risks associated with de-isolating other sources of energy, such as services, must not be underestimated. Some or all of the services such as steam, air, cooling water, condensate, hydraulic systems, nitrogen, and electrical systems associated with the plant may also have been isolated. Water hammer, steam/condensate hammer and maximum heating/cooling down rates must be carefully controlled. Regarding process isolation, checks must be made by the equipment owner to ensure that it is ready for de-isolation, and restoration of services must be carefully coordinated, particularly following a major turnaround. For example, electrical power may be required before air can be safely restored; and process air may be required before the boilers and turbines can be restarted. Under some circumstances, it may be necessary to rent mobile power, nitrogen, and/or air supplies before the facility is ready for de-isolation and startup.

11.6 CLEANLINESS

The de-isolation process may provide an opportunity to thoroughly check equipment for cleanliness and dryness, and to possibly locate tools or equipment that may have been overlooked. This is further detailed in Chapter 12.

11.7 BOLTS, FLANGES, AND GASKETS

A disturbed joints register should have been developed during the planning stage, as outlined in Section 9.2. This register allows all broken joints to be tracked throughout the maintenance period and ensures that they are properly inspected, and leak tested, prior to recommissioning. It is often necessary to leave some isolations in place for activities such as pressure/leak testing or catalyst reduction. Removing the final line blind or connection to a vessel may involve a "final" joint that has not been subjected to a pressure or leak test.

Bolt torquing standards for flanges and joints should be specified along with formal verification in the field. Critical joints may require bolt tensioning using specialist equipment. Management of gaskets and bolts is critical (including gasket standards regarding type, size, materials), and clear rules on replacement or re-use of old gaskets and bolts must be established. Pre-startup activities such as closure procedures for critical joints and retorquing requirements must be considered, as discussed in Section 12.4.

11.8 CHANGES TO ISOLATION BOUNDARY

Where changes are made to the energy isolation boundary, such as changing the position of valves and bleeds inside the isolation perimeter, the risks must be re-assessed and the equipment re-tagged and appropriately documented.

11.9 SUMMARY

This chapter addresses some of the issues involved in reversing the isolation plan, which can include risks similar to risks associated with the initial isolation. Clear identification of transfer of ownership, and detailed procedures that are understood by all parties, will ensure the isolation of the equipment can be safely reversed when needed.

12 RECOMMISSIONING

This chapter addresses preparing the equipment and associated process for return to service. Recommissioning occurs after the equipment has been removed from service, isolated from the various energy sources, decontaminated, serviced, repaired, modified, or possibly replaced if damaged beyond repair, and is now ready to be placed into operational service. Before re-introducing chemicals or materials to the process and restoring power or energy, a review of the operational readiness of the process must be conducted, the scope of which should be commensurate with the complexity of the equipment and the work that has been conducted, and the hazards of the process. The recommissioning review should be formally structured such that all aspects of re-introducing chemicals to the process have been thoroughly vetted by knowledgeable personnel who are responsible for the operation and maintenance of that process. Equipment must be tested to ensure that it can contain its chemical contents, that the functionality of safety systems and interlocks is re-established, and that the equipment can function within established operating parameters.

The nature and scale of recommissioning instructions should, again, be commensurate with the complexity and hazards associated with the process and can range from, or reference:

- Written Standard Operating Procedure (SOP) on routine activities such as a pump swap.
- Specific instructions, temporary SOP or checklist-type SOP that may include independent verification for complex/hazardous return to service.
- Generic training material, or specific training on reversing energy isolation.
- Walk The Line (WTL); a physical review inside equipment boundary (See 12.6)

Following a major turnaround, or where process or equipment modifications involving a Management of Change (MOC) procedure have been introduced, a formal Pre-Startup Safety Review (PSSR) is recommended. Some Regulators, including OSHA in the USA, mandate that a PSSR be conducted under such circumstances for processes that are covered by 29 CFR Standard 1910.119 - *Process Safety Management of Highly Hazardous Chemicals*[5] (CFR online; OSHA online). This is further detailed In Chapter 13 and Appendix D of this book, and in Appendix A of the CCPS book on *Guidelines for Performing Effective Pre-Startup Safety Reviews (*CCPS 2007a*)*, which includes several example PSSR checklists.

For other, more general maintenance activities that require line or equipment opening, an "operational readiness review" should be conducted to ensure that the recommissioning step is thoroughly and fully documented and approved. Operational Readiness is one of the 20 elements of the CCPS Risk Based Process Safety (RBPS) management approach (CCPS 2007b) and for high hazard processes will be very similar to a PSSR. An Operational Readiness Review may include a checklist, similar to one that is required as part of a PSSR. As a minimum, this checklist should verify that before starting up the equipment:

- A punch list of items that need to be completed before startup is made and verified in the field.
- All mechanical, electrical, instrument, and civil work is complete and signed off.
- Isolation and de-isolation plans are documented, completed, and signed off.
- Cleaning, purging, leak/pressure testing is documented, completed, and signed off.
- Commissioning and startup procedure is available and understood.
- Safety interlocks have been reinstated and tested.
- Process system alarms reset and tested.
- Walk the Line (See 12.6).
- The PPE requirements for recommissioning are prepared and understood.
- Post startup actions are detailed with timelines and responsibility for completion.

---

[5] OSHA 1910.119(i)(1): The employer shall perform a pre-startup safety review for new facilities and for modified facilities when the modification is significant enough to require a change in the process safety information.

A major cause of releases during re-commissioning is human error, such as inadvertently leaving drain/vent valves open. For higher risk recommissioning activities, particularly where procedures are complex, a formal assessment of human factors should be considered. This may involve a Safety Critical Task Analysis (SCTA) that examines each step to identify the potential for human error, the performance influencing factors that may increase the possibility of an error and mitigations that should be put in place to ensure the activities proceed to plan. Further guidance on SCTA is available from numerous sources including the Energy Institute (EI February 2020).

12.1 CHECKS FOR DEBRIS

Before proceeding with a recommissioning effort, but prior to closing and sealing equipment, a final physical check should be made for debris that may have been left inside the equipment. This could include items as varied as nuts, bolts, welding rods, and even rags and other materials used for leak/pressure testing. Some components such as non-return valves (check valves) or instruments may be fitted with plugs or stabilizers to protect the internals for shipping. The consequences of such items being left behind when the plant restarts could be disastrous (see also 11.1 and 11.2) or prevent the equipment from operating, as in Case Study 12-1. Where major work has been undertaken or potential consequences of leaving debris behind are high, formal closure reviews and approvals should be considered. This could include checks for debris, proper installation of internals, specified gaskets, and bolting requirements, which are formally documented and then signed off by authorized personnel.

Case Study 12-1 — Vent Line Blockage

Description
A new vent line, constructed from GRP (fiberglass) was installed on a low pressure storage tank. Once it was put into service, a pressure relief valve on the tank would lift whenever the tank was being filled. The vent line sizing was checked for capacity by calculation and found to be more than adequate. It was, however, eventually removed and an internal restriction was discovered in the vent line.

Comments
The GRP pipeline had been fabricated with two 90-degree bends, so it was not possible for someone to see through from one end of the pipe to the other. As part of the investigation, the team dropped a tennis ball through the line, but it did not come out of the other end. They cut the line open and a wooden mold that had accidentally been left behind by the fabricator was found in one of the bends. It had effectively formed a plug in the vent line.

Key Learning
Although this was an unusual incident, when new equipment is installed, it should be thoroughly checked to ensure that it can perform its required function, as part of the commissioning process. Similar issues arise due to line blinds and solid gaskets that are used for testing and not removed during the recommissioning.

12.2 CLEANING AND FLUSHING

Depending on the maintenance activity and the nature of the process, it may be necessary to clean and flush out the equipment. Oil and grease residues can cause issues with some of the process materials (e.g., oxygen, chlorine, peroxides, ammonium nitrate), and water or other flushing fluids may be incompatible with the process. Chemical cleaning may be required and in certain processes it may also be necessary to passivate the inside of the equipment using special fluids, e.g., a corrosion inhibitor. For food and pharmaceutical industries, further steps are required as part of Good Manufacturing Practice (GMP) requirements, including disinfection and sterilization stages. All of these issues should be considered in advance as part of the recommissioning plan.

12.3 PURGING

Returning equipment to service often necessitates various degrees of purging, which may be required before or after leak/pressure testing (or occasionally both). Purging requirements for specific equipment/systems must be specified and included in the recommissioning plan. The purpose of purging is usually to avoid incompatible substances in the process, and typically involves the removal of air/oxygen, removal of moisture to a specified dew point, and/or inert gas blanketing, such as nitrogen. There may be a requirement to use a warm or even hot purge gas to help with drying, in which case a detailed risk assessment is required to check for material issues within the process. Some processes, especially where even tiny amounts of a particular substance (including water) are a major issue, require the equipment to be subjected to full or partial vacuum prior to recommissioning. Again, this requires a detailed assessment to ensure that all of the vessels and pipes are designed to cope with the vacuum. In some rare cases, process gases are used for purging, which requires a completely different level of risk assessment, including external risks and environmental impacts.

Where temporary equipment is brought to the site to provide purge gas (such as a nitrogen bottle bank, air compressor, or ambient air heater) a risk assessment potentially including a temporary Management of Change (MOC) is required to ensure that the equipment cannot be accidentally over-pressurized or overheated, and that personnel are not at risk of burn injury (if, for example, pipes are not insulated).

Under some circumstances, it may be necessary to consider the composition of any mixed gases in a flare system, to ensure that a corrosive or explosive mixture is not inadvertently created. For example, purging a chemical stream rich in oxygen to a flare header that is being used for hydrocarbon flaring may create an explosive condition that could lead to an explosion within a flare header or flare stack.

The purge plan for the equipment should consider all connection points for draining and venting, low and high points in the equipment and process, and dead-end locations, as well as any instruments/valves/devices that have a complicated design that makes purging difficult to accomplish. In many situations, the connection points utilized for decommissioning the equipment can also be used for recommissioning.

The plan should also consider the rate of purging and any measurements that may be required (flow, temperature, pressure, composition) to achieve the desired outcome safely. When drying processes to remove water to a very low level are included in the plan, pressurizing, and depressurizing the process may be required to be repeated several times. Any measuring instrumentation that has not recently been tested or calibrated must be checked before use.

12.4 LEAK/PRESSURE TESTING

The requirement and scope of leak/pressure testing should be considered and thoroughly discussed at the initial planning stage, well before beginning recommissioning. Inspections such as NDE (Non-Destructive Examination), visual inspections and torque checks on bolts must be conducted prior to leak/pressure testing. For processes where the equipment is handling toxics, flammables, combustibles, corrosives, oxidizers, and regulated substances, leak testing is normally included in the recommissioning step. The testing protocol must be discussed with SMEs and documented, so that it meets the leak testing objectives, does not exceed maximum allowable working pressure (MAWP), and does not create a safety hazard for the personnel performing the test. The primary objective of the leak test is to identify any potential leaking points, open vents or drains, or missing piping components such as gaskets or plugs. Higher standards of leak testing are often required for commissioning a new plant.

In cases where the equipment has been modified or replaced and is governed by codes or standards (e.g., ASME B31.3 for process piping or ASME Section 8 for pressure vessels), a certified pressure test of the equipment may be required. The method and procedure for the test must be reviewed and approved by a Subject Matter Expert (SME) and must consider the risks created by the pressure test. For example, the use of compressible fluids such as air and nitrogen versus non-compressible fluids like water must be considered and discussed. See Case Study 12-2, where a single block isolation was not adequate for pneumatic pressure testing of piping connected to a tank (CCPS Beacon October 2007). Compressible fluids can pose a catastrophic consequence if the equipment ruptures, whereas non-compressible fluids normally pose lower consequences if a failure occurs.

Testing fluids should be compatible with the process fluids that will subsequently be introduced to the system. For example, water should not be used for testing equipment that is to operate in cryogenic service, unless there is a rigorous procedure for drying out the equipment. For low risk applications, it may be sufficient to allow "service testing" where the process fluid is used to test for leaks.

Case Study 12-2 — Vessel Failure During Pneumatic Pressure Test

Description

A plant was conducting a pneumatic pressure test on some pipework that was connected to a tank and was isolated via a single block valve. During the test, the tank suddenly failed at the base (photo, left), was launched into the air, and came to rest at the top of the plant (photo, right).

Comments

The single block valve was leaking by (passing), allowing compressed air into the tank, which was not fitted with a relief valve. The tank over-pressurized and failed at the base. The stored energy from the compressed air was sufficient to launch it skyward.

Key Learning(s)

- Pneumatic testing should be avoided due to the danger of stored energy release if there is a failure. Hydraulic testing involves much less stored energy.
- Isolation via a single block valve is not reliable. A line blind or physical disconnection and fitting a blank provides a higher standard of isolation.
- Emphasizes the need to keep non-essential personal away when equipment is being tested.

When the process is designed to contain highly toxic or hazardous materials and even molecule-sized leaks are a concern, a sensitive leak test may be required. This type of testing may utilize helium and require specialized leak detection equipment operated by personnel who are trained in its use. To ensure the sensitive leak test is accurate, a detailed checklist should be developed to ensure the correct ratio of components, the right steps are used for pressurizing the system, and the test locations and method are fully documented. Soap and bubble testing can also be used for leak detection when sensitive leak detection is not required. Soap and bubble testing is conducted by applying low pressure air or nitrogen to the process and a soap solution to the outside; leaks are observed by the formation of bubbles at the leak source.

Newer technologies for leak testing include acoustic leak detection, and acoustic imaging, which uses digital microphones and video coupled with intelligent systems that can localize leaks.

One of the challenges to creating an overall plan and effective leak/pressure testing protocol is to ensure that all potential leak points and final connection/disconnection points are included in the plan. The plan should identify any equipment that is to be excluded from the leak test, which may include rupture discs or rotating equipment. A disturbed joints register should have been developed as part of the original planning stages (See Section 9.2), which allows all broken joints to be tracked throughout the maintenance period and ensures that they are properly inspected and leak-tested prior to recommissioning. The final connection/disconnection points, such as a flange, often receive extra scrutiny since they are the last points subject to opening/closing the equipment. Some companies have a policy requiring that these critical joints include a signature panel affixed to the equipment so the fitter can verify/sign off that the work is completed.

If the applicable codes for the equipment require a hydrotest or pressurized liquid test, then the recommissioning plan may require replacement of gaskets due to incompatibility of the testing fluid and the process fluid. If this is the situation, either a compatible testing fluid must be used, or the recommissioning plan must include replacement of the gaskets and an additional alternative leak test. No matter what tests are performed, the leak/pressure testing results should be fully documented, reviewed, and signed off before returning the equipment and process to service, as unforeseen issues can lead to serious events, as demonstrated in Case Study 12-3.

Procedures may include a requirement to re-check joints and equipment after warm-up/startup since joint integrity may be reduced at higher temperatures and pressures.

Case Study 12-3 — Thermal Oil System Boil Over and Fire

Description

A new chemical plant being commissioned for the first time included a hot oil system. The system consisted of an elevated head tank, oil heater, and an oil circulation pump. As part of the testing protocol, workers were required to leak test the oil system with water, before introducing oil. Since any residual water could cause a problem, the commissioning plan also included a "hot-hold" stage in which the circulating oil temperature would be maintained at 100 °C for a period of time, to allow the water to boil off slowly. The workers did this, but when they subsequently increased the temperature, a sudden emission of hot oil from the head tank cascaded down through the plant and ignited, causing major damage to the facility, but no major injuries.

Comments

Some of the water was not driven off at the "hot hold" stage, possibly due to the system being at an elevated pressure, or perhaps a slug of water was being held behind a control valve. The residual water boiled suddenly, forcing the oil upwards and out through the vent of the head tank. The ignition source was not identified. The protocol was well-intended since a leak test had been done, and it was anticipated that only water would have been released. Nevertheless, the risk of a boil over may have been underestimated.

Key Learning

Commissioning and recommissioning procedures require detailed risk assessment, potentially including a HAZOP, to ensure all possible negative eventualities are considered and properly managed.

12.5 FUNCTIONAL CHECKS

Functional field tests are required on valves, instrumentation, safety shutdown systems and interlocks, rotating equipment, and specialty in-line monitoring and detection devices. This should be an integral part of the outage plan and records of their completion should be documented prior to recommissioning activities. The tests must be performed per established Inspection, Testing, and Preventive Maintenance (ITPM) procedures by appropriately trained personnel. The outcome of the functional tests should demonstrate that the equipment functions as designed, and that the test results are documented.

Functional testing of some equipment critical to safety such as pressure relief valves (PRVs) cannot be tested online. Therefore, recommissioning must ensure that the inspections and tests for those devices are up to date. A good practice is to create a master list of all equipment and devices associated with the process that are being returned to service and to utilize a checklist to document when each functional test is complete.

Case Study 12-4 — Process Safety Scrubber: Testing is Key!

Description

During a maintenance shutdown/turnaround of a chemical plant handling a highly hazardous chemical, the process safety scrubber was overhauled. The overhaul included replacing the internal spray nozzles that distribute the neutralization fluid down the scrubber. There were several sets of spray nozzles within the scrubber, so to achieve the design spray pattern and minimize backpressure in the scrubber, the nozzles had to be installed based on the original design. A 3rd party maintenance contractor completed the nozzle replacement task. Per company requirements, after installation an inspection of the nozzles and functional test was conducted using water instead of the neutralization fluid. Water was seen coming out the top of the scrubber.

Comments

One set of spray nozzles had been incorrectly installed. The facility was able to correct the error and retest the scrubber before recommissioning it to service. Had the scrubber been returned to service without the inspection and functional test using water, a chemical release to the atmosphere would have occurred instead, during restart of the process. Also, the entire chemical process would have required shutdown and the scrubber flushed and cleaned before repositioning the spray nozzles.

Key Learning

Checking functionality of critical safety systems prior to returning them to service must be included in the recommissioning steps.

12.6 WALK THE LINE (WTL)

A fundamental requirement of recommissioning equipment and its associated plant process, including utilities, is carrying out a physical field inspection. Keeping materials in the pipes is key to preventing incidents. Hazardous materials may be released if process equipment is restarted when it is not yet ready for operation, or personnel safety hazards may still need attention.

This could be for many reasons including:

- Open ends
- Missing plugs and blinds
- Incorrect valve position – wrong line-up of equipment
- Loose or incorrect size of bolts on equipment/flanges
- Verification of gasket installation and torquing requirements
- Vent and purge hoses still connected/vent lines that are left open
- Labelling incorrect or not present
- Instruments and transmitters not valved in
- Open electrical wiring and boxes
- Safety systems not recommissioned
- Safety barriers not re-installed
- Flammable or combustible materials (such as wooden scaffolding and debris) near hot surfaces

As part of the recommissioning activity, it is essential to Walk the Line, to physically check in the field that all is as it should be. Physically walking the line can be best assisted and accomplished with P&IDs and/or checklists that show the isolations, break-in points, maintenance activity and de-isolations. Any deviations from what was expected must be thoroughly investigated. (Note: See the detailed summary document of Walk the Line in Appendix B.)

Walk the Line can comprise a *secondary check* of what should have already been carried out during the recommissioning activity. The walkdown does not replace plant operation personnel following their standard operating procedures (SOPs) for restarting the process after a turnaround or scheduled maintenance

Walk the Line refers to an essential physical review of equipment to ensure operational readiness. Appendix B describes the Walk the Line conduct of operations management system model in a broader context including elements of management, operational, and engineering discipline. The philosophy behind Walk the Line considers that to successfully produce consistent results, there are elements of leadership identifying the right tools, operators executing tasks correctly with those tools, and technical people monitoring results and changing the tools for continuous improvement. This applies to physical review for recommissioning as a tool to prevent process safety events and should be considered as a part of intentional competency development for those doing the physical review.

A CCPS Process Safety Beacon article (CCPS Beacon 2015) on Operational Readiness/Walk the Line refers to the Piper Alpha incident in the UK North Sea in July 1988 that led to 167 fatalities and could have been avoided, had there been a better system for checking the readiness of the equipment. In this case a pump was restarted but, unbeknown to workers, a relief valve in the pump discharge had been removed for service and a blank had been loosely installed in its place. This led to a release of hydrocarbon condensate that subsequently ignited. See Case Study 11-1 for further details.

12.7 DELAYED RETURN TO SERVICE

Equipment sometimes remains out of service for extended time periods, on both a planned and unplanned basis. Many processes have in-line spare equipment such as a pump, reboiler, condenser, or compressor that was taken offline for maintenance and repair. Typical issues that may occur with equipment that has been out of service for some time include:

- Dew-point corrosion, particularly in refractory lined-equipment
- Microbiological induced corrosion (MIC). Austenitic Stainless steels (and other materials) are particularly susceptible to this type of corrosion if laid up with uncirculated, untreated water.
- Deterioration of bearings and seals

If the equipment is planned to be offline for an extended time period, extra precautions may be required to ensure that the integrity of the equipment is not adversely affected. This can include the following:

- Protecting the equipment from severe weather conditions such as high moisture, low temperature
- Maintaining seals and rotating parts
- Sealing off openings or connections
- Establishing and maintaining an inert gas concentration
- Monitoring for corrosion and deterioration.

When unplanned delays occur, the entire recommissioning plan may need to be reopened and re-applied. Other factors to consider during delayed recommissioning could be that plant process conditions may have changed, different personnel may now be involved in the repair work and decommissioning, or the equipment may have incurred some unanticipated mechanical integrity issues. Under these types of situations, a Management of Change (MOC) study should be conducted that would trigger a formal PSSR.

Case Study 12-5 — Spare Reboiler Explosion

Description

A chemical facility had two reboilers for the distillation process: one was online and the second was a spare reboiler that was isolated by two single valves—one on the inlet and another on the outlet. The reboiler in service would foul over time, requiring that it be taken offline for cleaning and the spare reboiler would be returned to service. During return to service of the spare reboiler, a catastrophic rupture of its shell was followed by a BLEVE, causing multiple fatalities.

Comments

Sometimes the spare reboiler would be offline for over a year. During the period offline, flammable vapors leaked past the single block valves into the spare reboiler and condensed, likely mixing with the nitrogen that was introduced after it was isolated. As it was being returned to service, the heating medium for the spare reboiler was recommissioned by opening the valves but the reboiler was still offline and thus isolated from the distillation column and its pressure relief device. The reboiler heated up, causing the liquid to expand and pressurize the shell of the reboiler, resulting in its catastrophic rupture.

Lessons Learned

There are multiple lessons resulting from this incident. However, the lessons associated with preparing equipment for maintenance and return to service include:

- Isolation valves can leak if chemicals are introduced into sections of the equipment or process that have been deemed decontaminated. Before returning equipment to service, the likelihood of leaking isolation valves and the presence of chemicals should be considered.
- Equipment that is isolated from service for an extended timeframe requires special consideration and a fully risk-assessed isolation and return to service plan; possibly a formal PSSR.

13 PRE-STARTUP SAFETY REVIEW

CCPS published an informative guideline book: *Performing Effective Pre-Start Safety Reviews,* in 2007 (CCPS 2007a). This chapter refers to that publication and provides additional material on matters relevant to returning to service equipment that has been offline for maintenance, repair, or modification.

13.1 PSSR AND RETURNING EQUIPMENT TO SERVICE

PSSR is commonly applied within chemical processes and facilities as the final check, initiated by an event, change, or new installation, prior to the return to service or place in service of equipment. The goal of the PSSR is to ensure that the modified or new process equipment is safe and ready for startup. Thus, the PSSR may be similar to the recommissioning activities and "operational readiness reviews" discussed in Chapter 12. Some facilities might use these terms interchangeably or other expressions such as "Pre Startup Audits," but several differences to note are as follows:

- As discussed in Chapter 2, PSSR is referenced and *required* by many regulators, and a comprehensive recommissioning procedure is a good management practice that will closely resemble a PSSR.
- PSSR is typically very formalized and utilizes a more structured checklist process.
- PSSR often marks a key point in time on a project schedule whereby responsibility may pass from one organization to another (such as from engineering to construction to operations)

The PSSR is intended to be a *formal review* of the manufacturing process to verify that changes and new equipment designs have been assessed thoroughly prior to the introduction of chemicals and energy. To ensure this, a PSSR policy/program should be established to include criteria for initiating a PSSR, responsibilities, workflow, and documentation tracking. Since there are many steps to complete during the PSSR, a workflow chart such as the typical PSSR workflow shown in Figure 13-1 (CCPS 2007a) could be helpful.

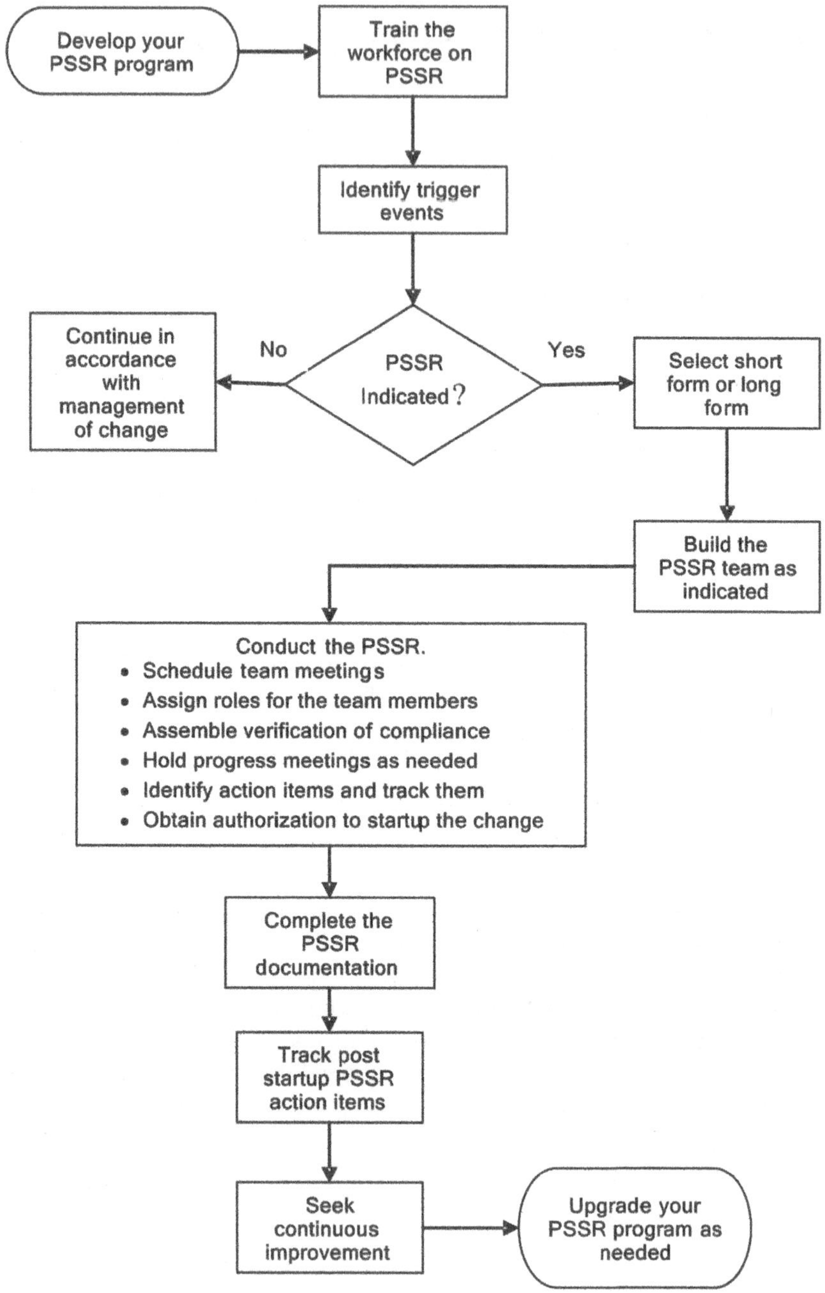

Figure 13-1 High Level PSSR Workflow Chart

As shown in the flowchart, the PSSR may be conducted using either a short form or a long form. The short form is typically an abbreviation of the long form and may be applied to changes that are very limited in scope, only require a few disciplines to approve, or are of temporary duration. A company's management policy for PSSR must include guidance on selecting the type of PSSR to apply.

In addition to the workflow shown in Figure 13-1, Table 13-1 from the same source (CCPS 2007a) provides several references to Global Examples of PSSR related documents and provides titles/descriptions of the documents and access to additional examples of PSSR guidance, (e.g., Marsh & McLennan 2016).

Table 13-1 Global Examples of Pre-Startup Safety Review Related Documents

ORGANIZATION	DOCUMENT
AIChE Center for Chemical Process Safety - 1989	Guidelines for Technical Management of Chemical Process Safety
American Chemistry Council (ACC) - September 1990	Responsible Care Code of Management Practices
Environmental Protection Agency (EPA) Rule 40 CFR part 68 - 1996	Risk Management Programs for Chemical Accidental Release Prevention
Health and Safety Executive (HSE) United Kingdom - 1984	A guide to the control of Industrial Major Accident Regulations, HS (R)21 (rev.)
ILO, International Labour Office (ILO) - Geneva, Switzerland - 1990	Prevention of major industrial accidents
International Organization for Standardization (ISO) - 1995	Petroleum and natural gas industries - Health, Safety and Environmental Management Systems
Occupational Safety and Health Administration (OSHA) Rule 29 CFR 1910.119 - May 26, 1992	Process Safety Management of Highly Hazardous Chemicals; Explosives and Blasting Agents; Final Rule
Organization for Economic Cooperation and Development - 1992	Guiding principles for chemical accident prevention and response
Official Journal of the European Communities - September 1985	Council Directive of 24 June 1982 on the Major-Accident Hazards of Certain Industrial Activities (the Seveso Directive)

ORGANIZATION	DOCUMENT
United Nations Environment Programme (UNEP) - 1996	Safety, health, and environmental management systems
World Bank - 1985	Manual of Industrial Hazard Assessment Techniques
Korean OSHA PSM standard, Republic of Korea -1996	Korea's Industrial Safety and Health Act - Article 20 Preparation of Safety and Health Management Regulations
Marsh & McLennan Companies (Marsh) - 2016	Pre-Startup Safety Review – Risk Engineering Position Paper – 06

13.2 PSSR OVERVIEW

The PSSR procedure requires a multi-disciplinary team of key stakeholders in the project, which might include the design engineer, process engineer, Management of Change (MOC) coordinator (if applicable), maintenance/reliability engineer, operations manager/supervisor, instrument engineer, safety specialist/representative, emergency response representative, process safety specialist, and trainer coordinator, among others.

Once authorized, the team prepares and reviews the checklist, inspects the site, reviews the documentation and training, inspects the facility, and provides a list of actions and a timescale for completion (e.g., before startup, within a week of startup, etc.). The checklist is completed and signed off, and any recommended actions are uploaded into a tracking system (sometimes referred to as a "punch list") to ensure the listed tasks are completed per the designated timescale. These are sometimes referred to as "Short" and "Long" Forms, based on the level of complexity, which is commensurate with the risk.

13.3 PSSR EXAMPLE: SHORT FORM

Table 13-2 shows an example of a risk-based questionnaire system. This short form is excerpted and reprinted from the CCPS book on *Guidelines for Performing Effective Pre-Startup Safety Reviews* (CCPS 2007a), with permission. A long form example, also excerpted from the same CCPS book, is included as Appendix D of this book. These examples and others can be completed in hard copy or electronic format. Many companies have electronic database systems for tracking the PSSR activities and the action items.

> **INSTRUCTIONS**: Complete the PSSR form with basic information for the review. If any special PSSR activities are indicated by the team, add additional pages as necessary. The PSSR team leader should track each applicable item as it is completed. When all items required prior to startup are complete, obtain the team member signatures. Transmit the PSSR completion to the party responsible for startup.

Table 13-2 Example Short Form for Lower Risk/Simple PSSR

<table>
<tr><th colspan="4">PRE-STARTUP SAFETY REVIEW - SHORT FORM</th></tr>
<tr><td>Date:</td><td colspan="3">PSSR Team Leader:</td></tr>
<tr><td colspan="4">Facility/Process Equipment Reviewed:</td></tr>
<tr><td colspan="4"></td></tr>
<tr><td>Type of Startup
(Check one)</td><td>New Construction:
☐</td><td colspan="2">Modified Process/Restart:
☐</td></tr>
<tr><td colspan="4">Recommendations: (Expand as needed. Attach relevant documents to this form)</td></tr>
<tr><td>Item No.</td><td>Description</td><td>Initials</td><td>Date Resolved</td></tr>
<tr><td></td><td></td><td></td><td></td></tr>
<tr><td></td><td></td><td></td><td></td></tr>
<tr><td colspan="4">PSSR Completion Summary: The following issues have been resolved and the undersigned believe the process/facility is ready for startup.</td></tr>
<tr><td colspan="4">• The construction and equipment meet design specifications.</td></tr>
<tr><td colspan="4">• Safety, operating, maintenance, and emergency procedures are in place and are adequate.</td></tr>
<tr><td colspan="4">• For new facilities, the initial process hazard analysis (PHA) has been performed and recommendations have been resolved.</td></tr>
<tr><td colspan="4">• Training of each employee involved in the operating process is complete.</td></tr>
<tr><td colspan="4">• Changes made to modify the process/facility have been reviewed and authorized by the facility management of change program.</td></tr>
</table>

Confirmation by PSSR Team Members		
Name/Title	**Signature**	**Date**

13.3.1 Post-turnaround Startup

Almost every chemical processing facility uses turnarounds to allow for major maintenance, detailed inspections, and equipment repair/refurbishing efforts. Typically, a turnaround involves a complete plant or unit shutdown and a significant work scope and coordination/pre-planning of multiple tasks. The work scope can include numerous maintenance requirements involving safely isolating and preparing equipment for initial opening, making changes to equipment, and returning equipment to service. Utilizing PSSR management tools during recommissioning can reduce the chances of missing or overlooking the actual status of critical equipment during startup. Case Study 13-1 further emphasizes the complementary relationship between PSSR and "operational readiness reviews" when returning equipment and processes to service.

Case Study 13-1 — Oil Pipe Rupture after Commissioning

Description

Shortly after commissioning, a flexible bellows on a thermal oil system ruptured and hot oil was sprayed over the process. It subsequently ignited, causing the destruction of a major part of the facility.

Comments

The circulating pump on the thermal oil system shut down for an unknown reason, but the low oil flow interlock failed to trip out the heater. The oil in the heater vaporized, leading to pressurization of the oil circuit, which failed at its weakest point. During earlier commissioning, problems had been encountered with the low oil flow trip, so workers temporarily bypassed it by linking it out using a short section of wire. The commissioning engineer had made a note of this in his personal notebook, but nothing else was documented and no further changes were made, leaving the system unprotected from loss of oil flow.

Key Learning (s)

- Changes made during commissioning and startup (even temporary ones) must be formally documented as part of a PSSR process.
- Interlocks must all be checked and tested as part of a PSSR.

13.3.2 Startup after Emergency Shutdown

Emergency shutdowns can occur for several reasons such as power interruptions, control system upsets, equipment failures, or acts of nature. If equipment failure is the reason, the failed item will need to be prepared for maintenance or replacement. Case Study 13-2 describes how relying on faulty readings on a new DCS led to a furnace explosion. The decommissioning steps outlined in this book include isolation, decontamination, and recommissioning. Depending on the extent of the maintenance or the need for a full replacement, an MOC and PSSR may be required. The PSSR can be applied during the recommissioning stage.

Case Study 13-2 — Fire Heater Explosion after Restart

Description

A fired heater tripped out due to an electrical fault and the operating team restarted it shortly afterwards. The process instrumentation had recently been upgraded so the gas valve positions could be seen on the DCS for the first time. The plant's startup procedure required a physical check to ensure that the gas valves were fully closed, as the first step in the sequence. The supervisor decided to use the DCS indication instead of going out and physically checking the valve positions. During the startup, a major explosion occurred, destroying the furnace, and injuring one person nearby.

Comments

The new system was faulty, and although it appeared on the DCS monitor that all of the valves were closed, one valve was actually slightly open, allowing gas into the firebox to create an explosive mixture before the ignition sequence started.

Key Learning(s)

- The new system was never intended to replace a physical check by the operators.

- There was insufficient training or understanding of the changes and no revised procedures were developed or communicated to the team.
- A formal PSSR, both for the DCS upgrade and for the furnace startup might have prevented the incident from occurring.

13.3.3 Restarting Mothballed Process or Equipment

Labelling a process or equipment as "mothballed", in a general sense, means that it has been out of service for an extended period and is placed in reserve or protective storage. The management procedure for mothballing should be carefully considered and detailed since the initial intent is to return the equipment or process to service at some future date. Typical steps to prepare the equipment for mothballing include isolation, decontamination, and special considerations to protect the equipment for an extended downtime. Recommissioning of the equipment and process reverses this procedure, and therefore may require additional safety and mechanical checks such as testing of pressure relief devices, valves, instrumentation, and trip systems, as well as checks on mechanical integrity. It may be necessary for the recommissioning team to establish whether there have there been any changes to the equipment during the mothball period, if the original isolation devices are still intact, and whether there have been other process changes that might affect the mothballed process or equipment. For these reasons, a PSSR is recommended during the recommissioning process.

13.3.4 Routine Maintenance and Minor Outages

For most short-term routine maintenance, a PSSR may not be necessary. However, management might decide that some additional maintenance checks, operational readiness review or a formal PSSR, would be beneficial to ensure the safety and reliability of the operation prior to a more formal recommissioning process. For example, extra scrutiny could be applied when replacing a catalyst, cleaning or rodding out heat exchangers, conducting maintenance of equipment critical to safety, modifying emergency backup power generators, or making changes to process control systems.

13.4 SUMMARY

Companies should provide management guidance for applying PSSRs when returning their equipment to service. A PSSR management tool can provide a structured approach to ensure that recommissioning of the equipment and the process have received a thorough check before they are returned to full service.

14 EQUIPMENT STARTUP

This chapter relates to considerations required when energy sources (e.g., process fluids and services) are systematically restored to the part of the process that was shut down for maintenance. For a routine startup, Standard Operating Procedures (SOPs) and checklists should be available. For non-routine startup, temporary SOPs might be written and again, a checklist approach should be adopted to ensure the correct sequence of events and verification takes place. The startup checklist can be included in the recommissioning plan, or specific checklists can be created for individual equipment and sections of the process. The primary objective of a checklist is to ensure that the recommissioning activities have been completed, the PSSR, if utilized was completed with sign offs, appropriate staffing is present, and personnel are aware of their startup responsibilities. Senior plant leadership will normally conduct a final review of the startup preparation and approve whether the startup can proceed or not.

14.1 PARTICIPATION AND COMMUNICATION

All participants, including field operators, control room operators, instrument and mechanical technicians, contractors, and plant management must have a common understanding of the various stages of the startup, their responsibilities, and the actions they should take.

Some facilities dedicate a suitably trained and competent individual (e.g., Process Safety Startup Coordinator) to oversee activities during pre-defined complex/hazardous startup activities. Duties could include ensuring that safeguards (both engineered and administrative controls) are available and operating as intended to prevent process safety incidents. The function is established to provide absolute control including stop work authority (SWA) but primarily to ensure procedures are followed and hazards are properly addressed.

Site safety and security personnel may provide support for the startup, particularly by reviewing and verifying that safe work permits are completed and issued, limiting access of non-essential personnel to the process unit and surrounding areas, and establishing an emergency response plan as appropriate. The plan may also include notification of adjacent populations both onsite and offsite. For particularly high-risk startups, it may be sensible to schedule it for when there is minimal population in the area.

Where startup is expected to last more than one shift, it is essential that a high quality handover to the next shift takes place due to the additional risks in this transient mode of operation. This handover is typically referred to as shift-to-shift communications. The status of the plant, procedures, deviations, work permits, and SIMOPs activities should be transferred in both written and verbal handovers. All personnel in charge should give an update on the specific areas they are responsible for, to ensure that the incoming shift is fully aligned on the status of the plant and the work to be accomplished.

Communication among these teams is paramount and a good public address or VHF radio system is essential. Communications should, in many cases, be maintained throughout the startup activity until the equipment is safely returned to service. Therefore, pre-checking the communication tools prior to commencing startup is highly recommended.

14.2 SPECIFIC STARTUP ISSUES

Some facilities that are starting up from cold may encounter specific issues that require special procedures. For example, there may be a requirement to override trip systems until certain process conditions are satisfied (pressure, temperature, flow, etc.). These should be documented in normal cold startup procedures, although additional requirements may also be required. If the overrides are manually initiated, safeguards to ensure that they are re-activated must be in place. The safeguards could include timers, computer alarms, or a checklist. Due to these many required checks and other factors, the potential for human errors is known to be higher during process restarts. High-risk startup activities can result in severe consequences— see Case Study 14-1 (CSB online 2011).

Some processes require special startup considerations due to known conditions that must be addressed. For example, process units that handle hydrogen (e.g., catalytic reformers) are susceptible to leaks from heat exchanges during startup. This is due to hydrogen pressure, the small size of the molecule (that makes it more leak-prone) and the cold temperature of the exchanger flanges. Typical procedures might involve directing steam hoses onto the leaking flanges until they come up to temperature and seal themselves. Historical or known specific startup conditions and potential consequences must be incorporated into the startup plan and checklist.

Additional Considerations for Startup:

- Retorquing of bolts/flanges from a cold startup.
- Heat up cycle time to minimize expansion stress.
- Replacement of gaskets that were damaged or contaminated during the pressure test or startup.
- Availability of specific leak detection devices.
- Technical representative of Vendor package available on standby.
- Extra PPE for personnel.
- Shift-to-Shift communication tool when startup involves more than one shift.
- Removal of physical barriers and platforms that were installed for equipment maintenance but are no longer required.
- Installation of additional barriers for the startup, if required.
- Relocation of non-critical personnel from the site and adjoining areas.
- Safety equipment such as fire extinguishers, hoses, ladders, emergency breathing equipment and masks are in place and checked.
- Emergency alarm and notification system operational.
- Backup emergency power equipment tested and ready.
- Notification of the startup to onsite personnel and to offsite facilities (if they could be affected).
- Consider siting issues including temporarily removing personnel housed in wooden trailers, etc.

Case Study 14-1 — Kleen Energy Natural Gas Explosion

Description

Part of the commissioning and startup phase of the Kleen Energy project involved cleaning of fuel gas piping. Workers were conducting a planned "gas blow," a standard practice where natural gas is forced through the piping at a high pressure and volume to remove debris that might be left inside the pipes. Natural gas and debris are subsequently released directly to the atmosphere. Workers were using natural gas at a pressure of approximately 650 pounds per square inch gauge (psig), which was being blown from an open-ended pipe between two large structures, known as heat recovery steam generators (HRSGs), in an area immediately south of the power generation building. There was a sudden large explosion and six individuals who were working within the power generation building were fatally injured.

Comments

The "gas blow" purging procedure with natural gas was considered a normal practice by utility companies. Agencies overseeing utility facilities referenced the procedure in their publications. The risk was considered low if the natural gas was blown to a safe location. This location, while outdoors, was congested by the surrounding power generation equipment. The vent pipe had been installed in a relatively horizontal orientation. Both the congested area and the orientation of the vent pipe likely affected the dispersion of the natural gas. Efforts had been made to eliminate or control potential ignition sources outside of the power generation building, but ignition sources remained outside and inside. The gas blows could have been self-igniting due to either the expelled debris creating a spark, or through static accumulation from the flow of the gas. Additionally, many ignition sources also existed inside the building: electrical power to the building was on, welders were actively working, and diesel-fueled heaters were running.

Approximately 150 workers were at the construction site the day of the explosion. Non-essential personnel had been restricted from the area immediately south of the main power generation building during the gas blows. However, more than 50 people were working inside the power generation building at the time of the explosion; only about 15 of those 50 people inside were involved in the natural gas blow activities. Five of the fatalities were involved with the natural gas blow activities and one was not.

A risk assessment had not been conducted for this activity, and the procedure was executed without the cleaning crew conducting a safety review.

On the day of the incident, the pipe cleaning crew did not have a safety meeting to specifically discuss the hazards of natural gas blows, nor did they receive and review the natural gas blow procedure.

Key Learning(s)

- Inherently safer alternatives to high-risk activities such as natural gas blowing should be considered.
- A formal risk assessment is required for commissioning activities, such as the natural gas blow procedure. Had one been conducted, it should have identified that there was the potential for an explosion due to the presence of a flammable mixture, multiple ignition sources, and a congested area.
- A safe work plan should be developed from the risk assessment, to include specific safeguards and should be communicated to all workers.
- Only essential personnel should be present for high-risk activities. One of the fatalities was not involved in the gas blow procedure.

14.3 STARTUP OF MOTHBALLED EQUIPMENT

For various reasons, facilities will remove equipment from service. The equipment may be purposed for salvage, resale, or "mothballing". In all cases, the equipment will need to be safely isolated and decommissioned as discussed in this book.

When equipment is designated as "mothballed", the intended purpose is for future re-use of that equipment. Mothballed plant equipment is sometimes left in place but is more often moved to a specific storage area for used equipment (aka, the "boneyard"). Some additional precautions that should be considered when mothballing equipment include:

- Document the condition of the equipment including any preventive maintenance concerns.
- Retain all design and inspection, testing, and preventive maintenance (ITPM) records.
- Label or tag the equipment.
- Create a key of mothballed equipment.
- Establish an ITPM schedule to check on any changes to the condition of the equipment.
- Re-use of the equipment requires an MoC as well as a commissioning plan.

Neglecting such precautions can result in serious consequences–as described in Case Study 14-2 (CCPS 2020, CCPS Beacon 2015).

Case Study 14-2 — ARCO Channelview Compressor Explosion

Description

Wastewater from a propylene oxide/styrene monomer (POSM) process was stored in a 900,000-gallon (3,400m^3) tank prior to disposal. Hazards included the presence of oxygen in the vapor space from degradation of POSM and a light hydrocarbon liquid layer on the surface of the wastewater in the tank. There was a nitrogen purge to the tank to keep the vapor space below the lower explosive limit (LEL) that was controlled by an oxygen analyzer and the tank was vented to a scrubber via an off-gas compressor.

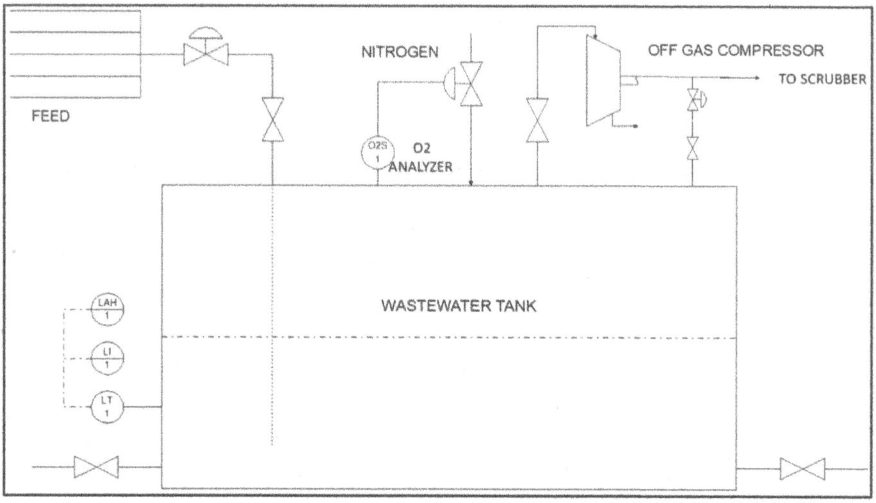

The off-gas compressor had been removed for maintenance and the venting of the tank was temporarily re-routed to atmosphere, as the tank continued in service. Approximately a week later, the nitrogen supply was turned off to allow for some modifications to the piping prior to reinstallation of the repaired compressor. About 34 hours after the nitrogen supply was isolated, the compressor had been reinstalled and an attempt was made to restart it when there was a violent explosion that destroyed the tank and flattened the surrounding area, resulting in 17 fatalities and $36 million of damage to the facility.

Comments

Unbeknown to the operators, the oxygen analyzer had become faulty, some 20 days before the incident. It was sending a false, low reading so the nitrogen flow had been reduced even before it was isolated the day before the incident. Operators assumed that the low reading was correct even during the recommissioning of the repaired vent gas compressor. The vapors were, in fact, above the LEL and they ignited when the compressor was restarted, probably due to mechanical friction in the machine. The flame spread back to the tank, causing the devastating explosion.

Key Learning (s)

- Hazard Analysis and Risk Assessment should have identified that the oxygen analyzer was part of a safety critical system. The safe operation relied on a single oxygen analyzer; redundancy is required on such a critical system.
- A formal risk assessment is required for recommissioning activities where equipment has been removed for maintenance.
- Any instrumentation that is required for the safe recommissioning should be tested/calibrated. This should include pressure gauges, analyzers and trip systems.

15 CONCLUSION

Many incidents have occurred as a result of poor or incomplete isolation, decontamination, and return to service activities, several of which are shared in this book. The purpose of this book is to provide practical guidelines for personnel in a variety of functions to help ensure that the shutdown, isolation, decontamination, and recommissioning of process facilities is conducted safely and efficiently.

There are multiple sources of energy that can lead to an accident or injury that must be considered. By starting with the identification and understanding of the hazards, and then conducting a risk assessment of the proposed work, potential safety issues can be determined, and appropriate mitigations put in place. This will include safe working practices and clear documentation, including instructions and checklists to ensure the work is carried out as specified.

Provision should be made for problems encountered during the isolation/de-isolation process and potential risks reviewed if there are changes, using a Management of Change process. Personnel must be trained for parts of the process that are relevant to their activities or area of responsibility. Various tools and techniques are available to help reduce the possibility of human error.

Equipment de-isolation is not usually a simple reversal of the isolation plan and various safety issues must be addressed as part of the early planning process, even before the facility or unit is shut down.

Examples of events that have occurred have been provided to demonstrate the types and scale of incidents that have occurred when these principles are not followed. It is hoped that this publication will contribute to a reduction in the number of such incidents in the future.

One final learning event is provided in Case Study 15-1, courtesy of the UK Energy Institute (EI Toolbox 2021). The event illustrates the importance of ensuring vent valves are directed to a predetermined safe location and taking extra precautions to verify that the isolation block valves are not passing chemicals.

Case Study 15-1 — Solvent Release from Bleed on a 'Double Block and Bleed' Valve

Description

A vapor recovery unit (VRU) located inside a building had been isolated for repairs. During routine rounds, the bleed valve was verified, and the operator confirmed there was no leakage.

Hours later, multiple lower explosive limit (LEL) alarms were activated in the VRU building, triggering a plant shutdown. The operator opened the discharge drum flare valve in an attempt to reduce the pressure and limit any release.

Solvent was released from the bleed on a 'double block and bleed' (DB&B) valve. Fortunately, the solvent did not ignite and there were no injuries. The bleeding valve was then closed and plugged to stop the release.

Comments

The investigation found that the downstream side of the DB&B required line pressure to energize the valve seal and make it effective. As line pressure dropped, the valve seal was lost and there was a flow out of the bleed on the DB&B arrangement.

The DB&B vent was left open and unattended, and left free to vent inside the building; this practice has been previously performed and accepted on site.

Elevated double block and bleed (DB&B) isolation valve and bleed valve

Key Learning(s)

- Venting inside a building is more hazardous than venting outside.
- Bleeds should be properly monitored and vented to a safe location (e.g., outdoors).
- DB&B valves must be of a correct design for the application. If they require pressure to create a seal, they may not be suitable.

APPENDIX A: WORK PERMIT

WORK PERMIT[6]

Safe Work Permit — Permit No.: 0000000

Any Change of the Scope Must Be Reviewed With The Custodian. This Permit May Then Need To Be Revised or Re-Issued.
THIS PERMIT MUST BE RE-EVALUATED IN THE EVENT OF A LOCAL ALARM IN THIS AREA OR ANY PLANT-WIDE ALARM

SCOPE

Date: ____________
Time Issued: ____________
Time Expired: ____________
Permit Extended Until: ____________
Work Order No: ____________ ☐ N/A
Work Order Reviewed: ☐ YES

Custodian Relief Change

Name	Time
____________	______ ☐ AM ☐ PM
____________	______ ☐ AM ☐ PM

Emergency Contacts:
Phone No.:
Custodian Channel:
Non-Custodian Group Channel: ______
Muster Location Discussion ☐ YES

Area: Equipment:
Non-Custodian Group Performing Work: Company: Craft:
Material Last In or Could Be Contained in Equipment:

Why Is Work Being Performed?

What Are Key Job Tasks/Steps? ____________

What Tools/Equipment/Utilities Will Be Used? ____________

How Has Equipment Been Prepared? (Drained, Cleared, Purged, Washed, Asphyxiant purge still on, etc.) ____________

The Scope of Work Includes Which of the Following Required Permits, Documentation, or Approvals – Check All That Apply

☐ Lock Tag and Try ☐ Line Break – SOP ☐ Excavation Permit ☐ Scaffold Inspection ☐ Other? ______
☐ Hot Work Permit ☐ CLWZ Permit ☐ Fire Impairment Permit ☐ Hot/Cold Bolt Checklist
☐ Confined Space Permit ☐ Fall Plan ☐ B/A Compressor Permit ☐ Blue Flag
☐ Line Break Permit ☐ Lift Plan ______ ☐ Automated Valve Use ☐ MOC # ______
☐ Line Break Permit Exempt Activity ☐ Energized Electrical Work Permit ☐ Food Safety Checklist ☐ Variance # ______
☐ Hydroblast Checklist

Risk Trigger: ☐ None ☐ Red ☐ Orange ☐ Yellow

HAZARDS

Chemical Hazards (List All)
☐ Flammable ☐ Eyes
☐ Toxic ☐ Eyes and Face
☐ Corrosive ☐ Hands
☐ Thermal ☐ Skin
☐ Inhalation ☐ Reactive
☐ Skin Irritant ☐ Asphyxiant
☐ Other: ______

Area Hazards (List All)
☐ Noise ☐ Suspended Load
☐ Lighting ☐ Electric Shock
☐ Congestion ☐ Arc/Flash Fire
☐ Heat/Cold ☐ Wind >25 mph
☐ Thermal Burn ☐ Standing Water
☐ Sharp Edges ☐ Ice/Snow
☐ Moving Parts ☐ Food Safe Area
☐ Magnetic Field ☐ Other? ______
☐ Radiation

Job Task Hazards (List All)
☐ Powered Mobile Equipment
☐ Lifting Equipment & Rigging
☐ Power Tools ☐ Hand Tools
☐ Sparks/Flame ☐ Fall > 4 FT
☐ Pinch Points ☐ Line of Fire
☐ Dust/Debris ☐ Ergonomics
☐ Slip/Trip/Fall ☐ Sprain/Strain
☐ Heavy Lifting >60# ☐ Other? ______

Protection Required For Hazards: Check The Appropriate Box To Specify Required Safeguards and PPE

☐ Goggles ☐ Face Shield ______ ☐ Chemical Hood ☐ Hearing Protection ☐ Double Hearing
☐ Chemical Gloves ______ ☐ Leather Gloves ☐ Dielectric Gloves ☐ Cut Resistant Gloves ☐ Impact Gloves
☐ Chemical Boots ☐ Dielectric Boots ☐ Chemical Suit ☐ FRC/Flash Suit ☐ Tyvek Suit
☐ Thermal/Steam Suit ☐ Long Sleeves ☐ Kevlar Sleeves ☐ Welding/Cutting PPE ☐ Dust Mask
☐ Half Mask Cartridge Respirator ☐ Full Face Cartridge Respirator ☐ Supplied Air –O2 Monitored ☐ Chlorine Escape Resp
☐ Harness ☐ Lanyard/SRL ☐ Anchor Point ☐ Positioning Device ☐ Warning Line
☐ Barricade Tags ☐ Barricade: ☐ Rigid ☐ Red Danger ☐ Yellow Caution ☐ Red/White Fall ☐ X-Ray
☐ Ladder Tie Off ☐ GFCI

Figure A-1. Safe Work Permit – Page 1 of 2

---

[6] Safe work permit (edited) courtesy of **Amy Gay, Occidental Chemical Corporation.** Principal Process Risk Engineer.

ISOLATION

Isolation: ☐ **Isolation of Job Scope Verified and Understood** ☐ **Not Applicable** (Scope Does Not Require Isolation)
☐ **Equipment / Line Not Cleared or Isolated** (Transfer of Control of Equipment provided to Non-Custodian)

Describe How The Point of Work Is Isolated: ______________________________

☐ Energy Sources Inside and Outside of Scope Identified ☐ All Isolation Devices Securely Isolated – Cannot Operate
☐ Field Verification of Isolations and Lockout Complete ☐ All Points of Work are Clearly Identified and Understood
Isolation of Energy That Cannot Be Locked Is Accomplished (Wheels Chocked, Jack Stands, Cribbing, etc.) ☐ Yes ☐ N/A

ENERGY

Verified Zero Energy? ☐ **Yes, Zero Energy Has Been Field Verified** ☐ **Not Applicable** (Scope Does Not Require Isolation)
☐ **No – Variance Required Before Proceeding** ☐ **No –** (Transfer of Control of Equipment provided to Non-Custodian)

Describe How You Verified Zero Energy: ______________________________

☐ Electrical Devices Verified With On/Off Switch, DCS, or Meter ☐ Interlocked Equipment Meter Tested ☐ N/A
☐ Equipment/Lines (Drained, Vented, Washed) Are Verified Clear ☐ Pressure Gauge or ☐ Open Bleed/Vent Verifies Zero Energy

AUTHORIZATION

PERMIT AUTHORIZATION

My signature certifies that I have personally completed verification of scope, hazard, process and energy isolation, a field verification* has been completed, and the work is safe to proceed. * ☐ N/A Field Verification Exempted Activity

Initial Permit Issuing Custodian – Print Name:__________ Signature: __________

Non-Custodian Craft Representative – Print Name:__________ Signature:__________ Time:______

Relieving Non-Custodian Craft Representative – Print Name: __________Signature:__________ Time:______

RETURN TO OPERATIONS

RETURN OF EQUIPMENT / WORK AREA – JOB COMPLETENESS (field verification required)

☐ **Complete** - The Work Scope Associated With This Permit Is Complete and Hazards have been Eliminated or Controlled (Guards Installed, Flange Guards, Piping/Instrumentation Intact, Housekeeping) and No New Hazards have been Created.

☐ **Ongoing** - The Work Associated With This Permit Is Not Complete
☐ Housekeeping is complete and there are no unsafe conditions related to the incomplete work.

Non-Custodian Craft Representative - Print Name:__________ Signature: __________ Time: _____ ☐ AM ☐ PM
Custodian - Print Name:__________ Signature: __________ Time: _____ ☐ AM ☐ PM

Authorized Workers

The Workers identified below have participated in a permit discussion and are authorized to work on this job.

Print Name	Signature	Print Name	Signature

Figure A-1. Safe Work Permit – Page 2 of 2

APPENDIX B: WALK THE LINE

WALK THE LINE

Walk the Line (as described below and shown in Figure B-1 by Jerry Forest LLC 2024) is a conduct of operations model consisting of tools designed to help manufacturing plant operators understand with 100% certainty where energy will flow to avoid line-up errors that might lead to process safety events.

CONDUCT OF OPERATIONS & THE WALK THE LINE MODEL

Conduct of operations is concerned with how work is done in a manufacturing unit to produce consistent results. The philosophy behind Walk the Line considers that to successfully produce consistent results, there are elements of leadership identifying the right tools, operators executing tasks correctly with those tools, and technical people monitoring results and changing the tools for continuous improvement. Each of these disciplines should be in place to ensure consistent performance and repeatable results.

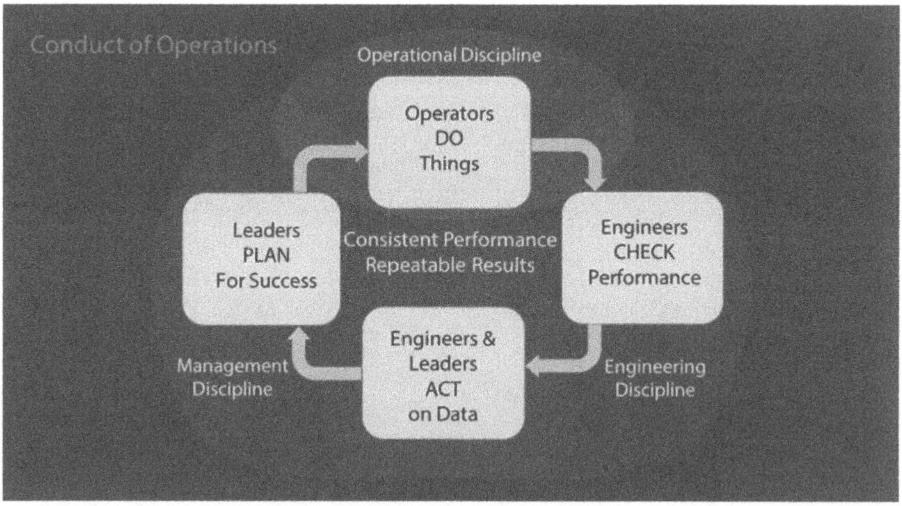

Figure B-1. Conduct of Operations Management System Model

MANAGEMENT DISCIPLINE

The Walk the Line model begins with leaders planning for consistency in performance by creating the operational discipline tools that enable operators to:

- Understand expectations and actions required to do their job (operators know what to do)
- Apply critical thinking when executing tasks (operators know how to do it)
- Create the culture that makes it easy to succeed (operators want to do their job)

Leaders create operational discipline tools, monitor results, and improve on those tools based on performance. They ensure operators are enabled to use the tools by training and reinforcement of their purpose.

OPERATIONAL DISCIPLINE

Operators have a finite set of routine tasks that are performed each day. They show up outfitted for work, make relief, prepare equipment for maintenance, issue permits, monitor the operating condition of the plant, take samples, execute tasks as instructed, and other activities specific to their role. To be successful, competency must be demonstrated for each task and tool used.

Operational Discipline tools that aid operators in understanding the current line-up of a plant include:

- Operational continuity tools – communication between and among shifts to ensure pertinent information regarding plant line-up is shared and understood by oncoming operators for the period since last worked
- Open bleeder management tools – how operators understand active open-ends in a process
- Line-up tools - how operators understand where energy will flow in a process when a change is made
- Operational readiness tools – how operators know where chemicals and energy will flow when returning any equipment to service

The most fundamental thing an operator does is line up energy and material in a process. If the operational discipline tools can't ensure with 100% certainty where the energy and material will flow, then an operator must physically walk the process to verify accuracy in line up.

ENGINEERING DISCIPLINE

One role of engineers and technical people is to monitor processes to optimize operation, and correct deviations before negative events occur. Processes are run by equipment and people. Equipment breaks, and people make errors. Engineering discipline related to Walk the Line includes monitoring the effectiveness of the operational discipline tools, understanding the causes of incidents, and making recommendations for modifying the tools to ensure optimum effectiveness. Engineers and technical people make recommendations to leaders for continuous improvement changes to the tools and usually take the lead in training and mentoring operators on the purpose of the changes to tools, and expectations regarding using them.

WALK THE LINE TOOLS

The conduct of operations model was developed after analyzing hundreds of incidents involving operator line-up error, and open bleed valve releases. Most of these incidents occurred during return to service or decontamination phases of transient operations. The tools have common elements of checklists, memory aids, and independent verification (where appropriate) that are specific to understanding line-up. The line-up aids are developed by cross-functional teams as part of management discipline. The purpose of each tool is documented and becomes part of intentional competency development for operators. The intended use is monitored, appropriate action taken on noted deficiencies, and immediate feedback is given regarding action or inaction. A summary of the more valuable tools in the Walk the Line toolbox follows. Note, this is not a comprehensive list.

Management Discipline Tools

Many of the incident causal factors included a lack of understanding expectations for line-up due to lack of training.

- Incident causal factor review
- Operational Discipline tool effectiveness review
- Training & mentoring plans on Operational Discipline tools

Operational Discipline Tools

Operational Discipline tools are designed to help operators understand the current operating state of a unit. Some common tools are listed.

Operational Continuity	Open Bleed Valve Management	Line-up Tools	Operational Readiness Tools
Shift handover	Open bleed valve tags	Line labeling	Commissioning
Shift instructions	Critical bleed valve program	Independent verification	P&ID walkdown
Shift change meetings	Plug program	Checklist SOPs/ instructions	Pressure/ soap testing
Operator shift notes	Bleed tag boards/ books	Material transfer considerations	MOC/ PSSR tags
Unit walk through			Startup HIRA reviews & PSSR
Evaluation rounds			
Bypass boards			

Engineering Discipline Tools

While many companies have the Operational Discipline tools listed, the causal factor analysis indicates that operators are not trained on the tools, use of the tools is not monitored, and many of the tools did not actually address the current state of the plant (i.e., how the unit line-up changes from shift to shift). Engineering Discipline addresses these concerns with the following tools:

- Evaluation of Operational Discipline tools
- Correcting Deviations
- Continuous Improvement

WALK THE LINE APPLICATION TO PREPARING EQUIPMENT FOR AND RETURNING FROM MAINTENANCE

Walk the Line was created primarily to address process safety incidents that occur when line-up changes are made to a process, and most of these changes occur when preparing equipment for maintenance and returning equipment to service from a downed state. Using Walk the Line Operational Discipline tools in conjunction with physical review of the plant will help minimize line-up errors leading to process safety events.

Understanding the Hazards in the Process – The chemical and physical hazards identified in maintenance preparation help to define decontamination plans commensurate with those hazards and define the Walk the Line tools used to ensure consistent results without process safety events.

Decontamination Plans - The complexity and detail of a decontamination plan is commensurate with risk. As with all Walk the Line tools, operators must be trained in the details and execution of decontamination plans and monitoring the success of those plans ensures consistent results without process safety events.

Energy Isolation – Many companies have their own rules and tools regarding energy isolation, and all involve physical verification of accuracy.

Return to Service – It is always the equipment owners' responsibility to ensure equipment is ready for service. Many Walk the Line incidents have causes that are related to open bleed valve changes within an isolation boundary, or failure to verify that equipment line-up or that equipment is ready to receive process. Each of these operational readiness causes are addressed by Walk the Line.

SUMMARY

A reliable plant running at steady state is usually a safe plant. Since normal operation usually accounts for over 90% of the operating state of a plant, transient operations can seem rare and deserve a more attentive approach to understanding risks and managing them.

The CSB has reported: *"37% of incidents investigated occurred prior to, during, or immediately following maintenance work."* (CSB 2017c). The key lessons from these incidents involve pre-planning and hazard identification for safe decontamination, proper energy isolation, and equipment readiness before return to service. Each activity involves people, and people can make mistakes.

Walk the Line aims to reduce human factor errors by analyzing the causes of these errors and providing tools to minimize the occurrence. It introduces a conduct of operations management system model to optimize success and sustainability.